GENIUS KITCHEN

ALSO BY MAX LUGAVERE

Genius Foods

The Genius Life

GENIUS KITCHEN

Over 100 Easy and Delicious Recipes
to Make Your Brain Sharp, Body Strong,
and Taste Buds Happy

Max Lugavere

Photographs by Eric Wolfinger

HARPER WAVE
An Imprint of HarperCollins*Publishers*

DEDICATED TO MY MOM, KATHY

CONTENTS

INTRODUCTION

In 2010, my mother, Kathy, began to complain of brain fog. I didn't think it was a big deal; she was fifty-eight, and I assumed her troubles were just a natural part of getting older. Many trips to many doctors' offices later, we found out what was really going on: the beginnings of a rare form of dementia. It was incurable, progressive, and with symptoms akin to having both Parkinson's disease and Alzheimer's disease at once.

In the months that followed, my mom began struggling to find her words or complete her thoughts. This wasn't easy for a woman who always had something to say. Eventually, even basic tasks became a challenge. One night she got trapped in the bathroom because she'd forgotten how to use the door handle. A knot would form in my stomach every time I'd see her struggle.

One thing my mother always loved was food, and she relished cooking for the family, but her growing impairment soon made ordinary tasks like using a knife or tending to an open flame dangerous. That's when I started to cook with her. She got to continue doing what she loved, and it helped make things feel normal when they were anything but.

A lifelong New Yorker and bon vivant, Mom also loved going out to eat, but her increasing fatigue made that hard as well. Still, I seized any opportunity to get her out of the house. On one particular late summer night, we had gone to dinner at one of our favorite neighborhood restaurants. It was busy but not loud, and the waitstaff was welcoming and kind to my mom.

I had spent the past few years studying nutrition, and I'd seize any opportunity to share my love for healthy eating with my mom—even when dining out. Whether or not she cared—I'd get an eye roll regardless—my mom was always happy to indulge me as we ate. On this night, however, something was off. She was quiet and had no appetite. In fact, she didn't eat a thing. I was worried.

The next day, I did what I had become so accustomed to doing: I took her to her doctor. Standing with her in the waiting room, I became suddenly aware of how decrepit she had gotten. She barely spoke and struggled to hold the saliva in her mouth. I took a tissue from the reception desk and gently wiped her lips. I held her hand as we walked into the doctor's office. He performed a routine exam and found nothing out of the ordinary.

Some days later, I was in Los Angeles for work when my brother called me. My mother's condition worsened while I was away, and he had taken her to the emergency room. Scared that something serious was in motion, I flew home and joined them, more anxious than I could ever remember being. I arrived at the hospital just as they had completed an MRI of her abdomen. The attending physicians gathered around me and my two brothers, who had left work early to be there.

Pancreatic cancer, they said.

As the doctors gave us the prognosis, I was in disbelief. How could this be? We—*she*—had already been through so much.

The following months were filled with every emotion a person can have: stress, anger, frustration, guilt, and, yes, even joy. Each moment with my mom during those last few months was at once beautiful and sad. We would try to feed her nourishing foods with what little appetite she had, but in the end, we sought to bring her joy with her favorite treats: strawberry shortcake pastries from Veniero's, a local New York baker.

When she could no longer feed herself, my brothers and I would each take turns, grateful for those remaining moments with her.

MY FOOD PHILOSOPHY

When my mom first became sick, I set out to learn as much as I could about the role food plays in our health. Since then, I've been invited to share my findings at conferences in my home country and abroad. I've had the opportunity to contribute to academic literature as well as teach healthcare professionals (including doctors!) about nutrition.[1] And I've published two books, *Genius Foods* and *The Genius Life*, which thousands around the world have used to achieve greater health and vitality.

I discovered that many of the chronic diseases people struggle with today—type 2 diabetes, cancer, heart disease, and dementia—begin years, if not decades, before signs and symptoms. In that time, what we eat (and how we live our lives) can influence our fate. But I also learned that food is a profound gift. In our busy world as we rush from obligation to obligation, we often forget to slow down and appreciate the meal in front of us. In this book, I will celebrate the act of eating in the hope that it brings joy and good health to your life.

My mother raised me to appreciate a wide variety of flavors. She enjoyed simple, no-frills cooking, but she also loved exposing me to new and daring cuisines from all over the world. On top of that, I've lived in New York, Miami (where I attended college), and Los Angeles, three of the world's great cultural melting pots. As a result, my recipes are designed to make a statement to your taste buds with bold and exciting, internationally influenced flavors and textures.

Though I began to cook when my mother could no longer cook herself, I realized that culinary literacy (like health literacy) had become yet another aspect of life that we had outsourced. I became passionate about creating healthy food that is delicious and easy to make, and have since had the privilege to learn from chefs all around the world, some of whom I've interviewed on my podcast, *The Genius Life*. And, having a teacher like my mom didn't hurt, either.

Most of my recipes will include low-starch vegetables, but I've also made use of some of my favorite tubers, like sweet potatoes, and even the occasional legume (white beans, for example). My dishes are lower in carbohydrates than more "mainstream" cookbooks, while offering you the freedom to increase or decrease the carbohydrate content to suit your needs. (When it comes to eating for good health, there is no "one-size-fits-all" diet.) All are nutrient-dense, gluten-free, and devoid of empty calories from added sugars and refined grains.

You'll also see me place the focus on nutrients like protein and fiber, important components of healthy, nourishing food. Eating like a "genius" means allowing your hunger mechanisms to naturally calibrate so that you feel genuinely satisfied once you've had enough. When you eat foods that are truly good for your body, they satiate innate hunger mechanisms in a way that fast food and packaged snacks just can't—and you'll discover why in the pages to come.

I wrote this book to be not just your new favorite cookbook, but a resource for your kitchen. In Part One, I'm going to do a deep dive into the science of healthy eating, breaking apart individual food components including cooking oils, dairy, fish, eggs, and even salt so that you can know how to leverage each food type for your best health yet. Then I'll go into my methodology for optimal cooking practices, as well as how to optimize your digestion. Part Two is where you'll find my recipes—a blend of main dishes, smaller dishes, and desserts.

My hope is that by reading this book—and enjoying my dishes with your friends and loved ones—you'll see food in a new way; able to bring joy and pleasure into your life without guilt. And you'll discover that you *can* eat foods that help you get fit and healthy without feeling like you're compromising on taste or giving up your favorite foods to do so.

Food is an intrinsic part of who we are; it is how we communicate, experience pleasure, show love and affection, bond with friends, and nourish ourselves. As I learned with my mom, our ability to appreciate food doesn't last forever. That's why each mouthful should be a celebration, and I'm grateful you've chosen to celebrate with me. Now let's get cooking.

HOW TO
EAT LIKE
A GENIUS

CHAPTER 1

STOCKING YOUR GENIUS KITCHEN

Welcome to the Genius Kitchen. My philosophy is simple: food should taste good, and make you feel good before, during, and after eating it. It should also help you to reduce your risk for major health ailments—not raise it. For years, I've combed the medical literature and have conversed with the top experts in the world, bridging the gap between the nutritional knowns and unknowns to bring you an achievable plan that brings joy and celebration to eating.

In this chapter, I'm going to lay out all the essential ingredients, from meat, to fish, to dairy, to salt, so that you know what to stock, and what to avoid, along with caveats to cater my recommendations to your unique biology. By the end, you'll have a strong sense of how to leverage food to look, feel, and age your best.

POWERFUL PLANTS

I was one of the lucky ones, born to a mom who knew how to prepare veggies. Whether it was the fresh, sweet Long Island corn she'd throw on the grill every summer, the savory salads she'd top with salt, pepper, garlic, and vinegar, or the cauliflower she'd season and roast in the oven, I never had an issue eating veggies thanks to the way my mother cooked them. In fact, she maintained a vegetable garden during my teenage years, growing tomatoes, zucchini, and various lettuces that we'd enjoy all year long.

Fruits and veggies take up permanent residence in the Genius Kitchen, and incorporating them into your diet promotes health in a number of different ways. For one, they're incredibly satiating from their high fiber and water content, but because they have low calorie density, you can fill up on them while ingesting relatively few calories. Plants also provide certain essential nutrients, like vitamin C, which are hard to get from other sources. And they provide a bevy of unique *non*essential nutrients that are thought to be beneficial to human health.

Fiber is the perfect example of a nutrient that, while nonessential, seems to make life better. After ingestion, it expands in your stomach, giving you a feeling of fullness. Journeying south, it helps sweep up toxins, keep certain hormones in check (estrogen, for instance), and even helps purge cholesterol-carrying LDL particles that your liver has sent for disposal (see page 33 to discover how). Then, fiber is what feeds the bacteria that live at the latter end of your digestive tract which churn out powerful anti-inflammatory chemicals as a form of thanks.* Perhaps this is why fiber consumption has been consistently associated with healthy aging (lower rates of cardiovascular and neurodegenerative disease, for example) and lower risk of certain cancers (e.g., breast).[2]

Plants contain other compounds that benefit us in roundabout ways.

---

* More accurately, these beneficial chemicals are a form of bacterial poop. But, hey, this is a cookbook after all.

Plants don't want to be eaten, so they develop chemicals meant to deter predators. In us, many of these compounds are not only safe, they stimulate a protective response—think of it as the biological equivalent of the Defense Production Act. Some "beneficial" plant toxins include sulphoraphane, produced when you chew raw cruciferous veggies like broccoli, and compounds called polyphenols, which are abundant in fruits, veggies, herbs, spices, and even coffee and tea.

Plants grown under harsh conditions tend to generate more of these defense compounds, and when you eat such plants, their "antifragility" gets passed down to you—a powerful example of the symbiosis of all living things. For instance, olives produce oleocanthal, which imparts anti-inflammatory benefits (more on this on page 37), and olive trees grown in harsher climates tend to produce more oleocanthal. Any wild plant will possess more vigor than its farmed counterpart, and organically grown produce tends to possess more than conventional.

These plant defense compounds aren't just good for you; they're good for your gut bacteria, too. When you consume fruits and veggies (or drink tea, coffee, or wine) your resident microbes, which thrive on such foods, release chemicals called *metabolites*. One such metabolite currently under investigation is urolithin-A, which is produced when we consume pomegranate. This nifty compound has been demonstrated to play a powerful role in neuroprotection—that is, protecting your brain from aging and decay.[3]

There are countless other such compounds, and science is just beginning to understand their effects. Generally found in higher amounts in low-sugar fruits and veggies, these compounds usually have strong flavors and pungent aromas; think curcumin in turmeric, the flavanols in dark chocolate, or allicin in freshly chopped garlic. You'll be happy to know that many of the recipes I've crafted for you utilize such ingredients, and that consuming them is consistently related to better health.[4]

RELYING ON PLANTS FOR ESSENTIAL MINERALS?

When it comes to essential nutrients—namely minerals like iron, magnesium, calcium, and zinc—there is some concern over their bioavailability in plants. While some plants do contain high levels of essential minerals (calcium in broccoli, for example, or zinc in lentils), they also contain compounds that act as "anti-nutrients." For instance, grains and legumes contain phytic acid, and dark leafy greens contain oxalic acid, both of which inhibit mineral absorption. Ultimately, in the context of an omnivorous diet—that is, a diet that also includes foods like beef, chicken, fish, and eggs—there's no need to worry, but these compounds can pose real deficiency risks for vegan and vegetarian populations. For added insurance, you can soak your grains and legumes in water, then drain them, before cooking (four hours of soaking reduces almost 80 percent of phytate content and seventeen hours removes phytates almost completely), or simply cook your veggies, which greatly reduces their anti-nutrients.

Finally, plant *pigments* have been demonstrated to play a powerful role in good health and performance. Of note, we find anthocyanins, which support gut and brain health, in purple foods like blueberries, purple potatoes, blue corn, and red onions. In foods with yellow undertones like kale, spinach, and Swiss chard, we find lutein and zeaxanthin, which help prevent age-related macular degeneration and cognitive aging. And in green foods, there's chlorophyll, a magnesium-rich pigment that helps reduce the absorption of food-borne toxins.[5]

With so much going for them, it's no wonder the research on fruits and veggies is so positive. We can see studies from animal models all the way up to observational and clinical trial data showing benefit.[6] Of course, everyone will be different, and your tolerance to individual plants and overall fiber amounts will vary. But by eating the dishes I've created for you, which incorporate a range of colorful, fibrous veggies and whole fruits (along with nutrient-dense animal products like beef, chicken, fish, and eggs), you'll be able to reap the benefits seen so frequently in long-lived people.

THE DOWNSIDE OF PLANTS?

When you peruse your local supermarket, what do you see? Fruits, vegetables, meat, fish, and aisle upon aisle of shelf-stable packaged foods. Whether it's cookies, cakes, pastas, breads, chips, cereals, meat substitutes, or "vegetable" oils, the vast majority of food-like products you'll find in your supermarket are plant-based. Government bureaucrats, the food industry, and celebrities alike continue to promote that some versions of these products, which are cheap to make and contain high profit margins, should constitute a significant part of our diets. Unfortunately, these products are highly processed and often nutritionally bereft, made from a convoluted slurry of unhealthful oils and pure starch harvested from the energy-collecting endosperm of plants like wheat, rice, or corn. These types of foods now comprise 60 percent of the calories we ingest every day and encourage a whole host of issues, including weight gain and metabolic dysfunction (see page 27 to understand why). While "whole food" plants like carrots, broccoli, olives, apples, avocados, and dark leafy greens are very good for you, be weary of "plant-based" ultra-processed foods—even if they're marketed as healthful.

When it comes to purchasing, do you opt for organic or conventional? Frozen, canned, or fresh? The answer is: it depends on your accessibility and budget! The research on farming systems is difficult to parse, being mired in the trenches of billions of dollars' worth of corporate interest. What we currently know is that while there is little nutritional value of organic over conventional, organic produce may have higher levels of beneficial phytochemicals (plant defense compounds, mentioned earlier), since their defenses against pests haven't been outsourced to man-made chemicals.

If organic falls comfortably within means, I recommend opting for that in certain circumstances. Prioritize organic especially for produce where you eat the skin or peel, such as apples and berries, and dark leafy greens and cruciferous vegetables (broccoli and Brussels sprouts, for instance). This will also reduce your exposure to synthetic, petroleum-based pesticides. There is no need to purchase organic lemons, avocados, bananas, and the like, because you typically remove the exposed skin or rind prior to eating.

When deciding between fresh, frozen, or canned, opt for fresh as much as possible. However, frozen can be a great way to economize and reduce food waste (I've provided more tips to save money and curb food waste on page 78). Try to minimize the consumption of canned veggies, because they usually have high levels of sodium added (more on this on page 51) and may leach hazardous, hormone-disrupting chemicals from the cans' plastic inner linings.

Here is a guide to know which kinds of plants to stock your Genius Kitchen with.

AVOID	GOOD	BEST
Refined grain flours. Refined sugar (e.g., cane, date sugar) and sugary syrups (i.e., agave). Ultra-processed "plant-based" foods.	Canned produce. White starchy tubers. Higher sugar fruits (e.g., mango, pineapple). Nut or legume (e.g., bean, lentil) flour–based foods.	Fresh or frozen whole food, low-sugar fruits (e.g., berries, olives, citrus, cacao, avocados), fibrous veggies (e.g., broccoli, cauliflower, spinach), non-white tubers (purple potatoes or sweet potatoes), nuts, seeds, legumes.

No matter your choices, plants are the ultimate "biohackers"; they interface with our biology in unpredictable ways that make us more robust and resilient. In a way, exposure to all of plants' myriad phytochemicals—foreign to our bodies—forces us to adapt and become stronger. But to be truly strong, we should consider animal products, the focus of the following section, which help us in other ways—by supporting our brains, musculature, and nearly every other aspect of our bodies.

MIGHTY MEAT

To me, animal protein means salty, unctuous, nutritious bliss. Whether it's a well-marbled rib eye, a nicely charred chicken thigh, fall-off-the-bone pork ribs, or even a simply grilled burger patty, when it comes to filling you up and nourishing your body and mind, it's hard to compete with a piece of well-prepared meat. And as satisfying as it is to your taste buds, it's equally satisfying to your body, containing the highest quality protein and other nutrients that nature has to offer.

I grew up in a house that was divided on meat—perhaps you did, too. My father enjoyed cooking it, and my brothers and I liked eating it. But my mother was a lifelong meat abstainer, consuming only certain types—usually lean poultry or fish—in strict moderation. My mom was a health-conscious eater and a lifelong animal rights advocate, which influenced her decision-making process at every meal. Considering meat had been demonized for so many years from both a health and animal welfare standpoint, it made sense at the time that she chose to avoid it.

Unfortunately, the advice that shaped my mom's diet and that continues to inform millions like her was pseudoscience: half-truths or mistruths from agenda-driven organizations and celebrity doctors, corporate profiteers, and mainstream health media. This messaging continues today, influencing

suggestible people to shun one of the widest and most important categories of nutritious food there is: meat. Meanwhile, we're getting fatter and sicker, while the business of ultra–high margin processed meat alternatives—the equivalent of human pet food—is growing hand over fist.

Let's take a step back. When you look at most of the meat produced in the United States, it comes from what are called concentrated animal feeding operations (CAFOs for short). Also known as factory farms, these systems are cruel to the animals. They're depicted in images you try to avoid when scrolling your Facebook newsfeed; the abhorrent hangar-sized barns where animals are stuffed in overcrowded pens, shackled, force-fed, covered in filth, and abused by underpaid workers. It's merciless and inhumane, and once seen can't be unseen. Rightfully so.

But alternatives exist. Today, consumers have access to farms that produce beef and other types of meat on a much smaller scale. On these farms, animals are allowed to roam on pastures for their entire lives, grazing and fertilizing the land. They consume their natural diets—not grain and other junk—transforming low-quality plants like weeds and grasses (which are useless to humans) to create powerful nutrition. The animals are vastly healthier and happier, and they ultimately experience "one bad day"—not the months of torture endured by feedlot animals.

In CAFO systems, younger cows consume grass but spend the final portion of their lives feeding on grain. Grain production is water intensive, requires irrigation, and is usually the result of a farming practice called monocropping, which damages our soil and yields higher levels of atmospheric carbon. Grass-finished cows, on the other hand, graze without the need for irrigated water. And by enriching the soil with their hooves and manure, they can actually cause the soil to *sequester* carbon, thus creating a net positive effect on climate change. This is known as a regenerative farming system.

Eating the right kinds of meat can also reduce animal suffering. Large-scale tilling and harvesting of crops causes the death of innumerable farmland critters like rabbits, squirrels, field mice, and insects—not to mention the wild birds and sea life harmed from herbicide spraying and runoff. "We know that animals are harmed in plant production," write Bob Fischer and Andy Lamey in the *Journal of Agricultural and Environmental Ethics*. While it's currently impossible to calculate the true cost to life, Fischer and Lamey offer a ballpark figure of at least 7.3 billion animal lives lost per year for industrial plant agriculture—in the same neighborhood as animal production.[7]

As you can see, "Meat is bad for the environment!" and "Meat is cruel!" are half-truths. Yes, 97 percent of our beef today comes from CAFO systems, while 3 percent comes from 100 percent grass-fed farms. But to partake in any aspect of modern life means complicity in a complicated web of industrial processes, the consequences of which are not always obvious. My hope is that you challenge the status quo by supporting regenerative farms. And while grass-fed and finished beef and pastured chicken can be more costly, it's worth it for the environmental and animal welfare, not to mention the benefits to your health. Whatever your budget, here is a continuum to guide your meat purchases.

AVOID	GOOD	BETTER	BEST
Fried or breaded meats. Fast food.	Unprocessed or minimally processed meats.	Pasture-raised. This is a great option for chicken, lamb, and pork.	Organic. 100% grass-fed, grass-finished. Free-range for chicken. Certified humane.

As I've mentioned, animals have an incredible ability to "upcycle" plants that are inedible to humans and turn them into highly bioavailable nutrients for us and our families. Protein is one of these nutrients, and animal protein is—without controversy—the highest quality protein available. We can get protein from plants, but generally speaking it is not as concentrated, or digestible, as it is in meat.* For instance, to get the same quantity of protein as one four-ounce chicken breast, you'd need to consume five cups of rice and beans, containing an additional 180 grams of carbohydrates and 25 grams of fiber, and the quality of protein is lower to boot.

There are many benefits to eating more protein. For one, it is highly satiating. Well before government dietary guidelines, our bodies evolved a way of encouraging us to pursue important nutrients. When you consume a rich source of protein, you satisfy receptors in your tongue, stomach, and brain that control hunger. For this reason, it's practically impossible to overconsume protein—it makes you feel so full! If you're hungry, reach for higher-protein foods and watch that craving dissipate.

Protein also helps to grow and maintain your lean muscle tissue, which is important for your strength, mobility, and health as you age. In fact, your

---

* This is according to the Digestible Indispensable Amino Acid Score (DIAAS), a current gold standard method of ranking protein digestibility used by the Food and Agriculture Organization. The DIAAS for animal proteins all land higher than 1, which is the ideal score for a food. The score for cooked rice is .6, and for beans it is .59.

muscles collectively make up a hugely underappreciated organ that may as well be considered a fountain of youth. They soak up sugar from your blood (chronically high blood sugar accelerates aging), and they even release compounds called *myokines,* which can boost bone and brain health. Even a small increase in daily protein intake can help maintain or grow new muscle tissue, which should be one of your main anti-aging strategies. Perhaps this is why adults over the age of sixty-five who consume more protein have greater longevity and reduced cancer risk.[8]

PROTEIN QUALITY AND MUSCLE MAINTENANCE

Protein is comprised of amino acids, and a protein is considered "whole" only when it contains all of the nine essential amino acids. Each amino acid serves a different purpose in the body. One amino acid, leucine, is particularly important for building and maintaining muscle mass, which is crucial to aging well. Gram-for-gram, plant protein contains less leucine than animal proteins like chicken, beef, or fish. For example, you'd need to consume 42 grams of protein from quinoa to get the same leucine content as 23 grams of protein from whey (a dairy protein). Vegan or vegetarian? You can pick up the slack with protein supplements (research suggests that .7 to 1 gram per pound of lean or goal bodyweight is ideal to support muscle growth and maintenance in exercising individuals).[9] One word of caution, however: many processed plant protein supplements harbor heavy metals.

One type of protein found in animal tissue that you won't find in the plant kingdom is collagen. Collagen is the most abundant protein in the body—it's the glue that holds it together while also providing elasticity to structures like your skin and arteries. Unfortunately, collagen production declines as we age, but by consuming more collagenous tissue (and vitamin C), you can actually increase your body's own production of it, improving your hair, skin, nails, joints, bones, and possibly even the rate at which you heal.[*]

Animal parts under heavy load or with wide ranges of motion such as shoulders, legs, and bones are rich in collagen—a chicken leg, for example, contains up to four times the collagen content of a chicken breast. Collagen is not particularly pleasant to eat in its raw state; if you've ever bitten into an undercooked, tendonous chicken leg, you know that firsthand. But if you cook those legs low and slow, all that delicious, nourishing collagen melts down into gelatin and becomes butter-soft. You'll find collagen in my Insanely Crispy Gluten-Free Buffalo Chicken Wings (page 194), Perfect Ribs (page 202), Brisket Chili (page 233), and Bone Broth Beef Stew with Purple Sweet Potatoes (page 191), among others.

Protein isn't the only nutrient found in meat, of course—far from it! In fact, meat is a valuable source of other essential nutrients, particularly nutrients whose intakes are often lower than what is recommended.[10] Take beef: it's an incredible source of vitamin E, iron, zinc, and creatine—a natural substance that supports your energy levels. What's most impressive about the nutrients found in beef and other types of meat is that they are *identical* to the form your body needs. Certain plant nutrients, for example, must endure complex biochemical transformation before your body can utilize them, and these processes differ in their efficiency and efficacy from person to person. Not so with nutrients derived from meat, which are easily recognized by your body and brain.

---

[*] Collagen synthesis also relies on vitamin C, which is found in fresh fruits and veggies. In fact, scurvy, which is caused by vitamin C deficiency, is a pathological inability to synthesize collagen. More vitamin C, with the inclusion of dietary collagen . . . a potent "recipe" for a youthful body!

Most of the meat we consume today (and most of the meat used in my recipes) comes from the skeletal muscle of animals. These are the muscles that animals use to move around with. But organs, like heart and liver, and bones are particularly rich in a number of important nutrients that are either lacking or less abundant in skeletal muscle. For example, heart is loaded with CoQ_{10}, an antioxidant that also aids energy production. Liver is packed with vitamin A, B vitamins, and copper. And bones are rich in minerals like calcium and magnesium as well as collagen. For picky modern humans, however, organ meats present a significant barrier to entry: funky tastes and textures. The key is to cook and season them well. With the exception of liver, most organ meats are rich in connective tissue (up to three times as much as skeletal muscle) and benefit from low and slow cooking. I love Bangin' Liver (page 192), which singlehandedly converted me to a liver lover. Bone broth, utilizing the bones of the animal, is another great dish (see page 137 for a simple recipe).

I've created many recipes for you that utilize animal proteins. I recommend dedicating half of your plate to protein; so if you're cooking a beef dish (a steak, for example), a perfect complement would be my Kelp Noodles with Brazil Nut Pesto (page 132). If you're making a stew that includes both protein and veggies in equal proportion, there's no need for additional produce, though you may appreciate a non-meat side dish. Feel free to mix and match to create your perfect meal!

FABULOUS FISH

Fish holds an interesting place in the American diet. It's either consumed as an indulgence exclusively in restaurants or reduced to junk food in the form of fried fish sandwiches, supermarket-bought fish sticks, or mayo-laden tuna salad. The reason for this? Most people simply don't know how to prepare fish at home. As a protein, it doesn't possess the vigor of beef, chicken, or pork, making it easy to overcook. But preparing fish the right way can be easy and rewarding to both your palate and your health.

When it comes to choosing fish, you always want fresh fish, or fish that was immediately frozen upon catching. It's only fish that is beginning to spoil that has that distinctive fish-market smell, resulting from the bacterial break-

down of trimethylamine oxide (TMAO), a compound naturally occurring in fish, to trimethylamine (TMA). Fresh fish should never smell overwhelmingly fishy. In fact, fresh fish often smells like crushed green leaves, with hints of ocean or lake water. Fillets should have a brilliant, glossy appearance, free of brown edges.

When it comes to cooking fish, there are numerous ways to get the job done while yielding delicious, moist, and flaky flesh. You can poach it, you can fry it, you can roast it, and you can grill it. Everyone's taste will be different, but I recommend medium to medium-rare when cooking fish. And while there's always a risk of contracting a food-borne illness when consuming undercooked meat, the risk from cooking fresh fish to a nice medium temperature is low, especially if the surface—where bacteria like to congregate—is thoroughly cooked.

Fish consumption imparts numerous health benefits, many of which stem from its fat. Fish, like any animal, need their cell membranes to stay flexible, but because the deep seas are cold, fish fat is predominantly comprised of polyunsaturated fats, which stay liquid at cooler temperatures. This allows them to stay agile when cold. (Beef, for example, has a higher proportion of saturated fat, which would turn rigid at oceanic temperatures—more on fats on page 27!) The colder the water, the oilier the fish, and cold-water fish fats are among the most precious we can ingest. They are the best source of omega-3 fatty acids EPA and DHA that support brain and cardiovascular health throughout life.

Among cold-water fish, salmon is the most commonly consumed today, but sardines are also loaded with omega-3s, can be found in most supermarkets, and are inexpensive. Tuna is also a great source of protein, though lower in omega-3 content. While all seafood today contains trace levels of mercury, a toxic heavy metal, tuna contains higher levels than salmon or sardines. Though mercury can be dangerous when we are exposed to it through industrial pollution, in tuna and other deep-sea fish it is packaged with an equal (or greater) proportion of selenium, making the fish harmless.[11] Still, moderate tuna consumption may be wise, especially for people who are young, pregnant, or breastfeeding.

In general, smaller, nonpredatory fish like salmon and sardines also contain fewer environmental pollutants than larger fish (i.e., tuna), which accumulate the toxins that they consume. But when industrially farmed, even healthful fish like salmon can store toxins that enter their bodies through the artificial diets they are fed. Some of these persistent organic pollutants include dioxins and polychlorinated biphenyls, or PCBs, which are linked to reproductive and developmental problems and cancer. This is why you should opt for wild, which contains markedly lower levels of these pollutants, or seek out responsibly farmed fish as an alternative.

If you can't find or afford wild, should you eat farmed fish? Absolutely. Like land agriculture, industrial aquaculture has made a number of mistakes over the past century as it has rapidly scaled up to feed a growing population. It is continually improving, however, with regions like Alaska and Norway leading the way. Now, 65 percent of the globe's fisheries are executed in a sustainable way, and this accounts for almost 80 percent of the fish that we eat, according to chef, TED speaker, and sustainable seafood expert Barton Seaver. Part of this has to do with the fact that the major supermarkets have adopted vigorous and continually evolving sustainability policies. Therefore, buying from a retailer that you trust can help ensure you're getting fish you can feel good about. Wild or farmed, just eat fish.

Consuming any product of the industrial food chain—even wild-caught fish—has drawbacks, but the benefits of fish consumption (even farmed salmon and tuna) do seem to outweigh the risks. In one study of older adults, eating seafood twice a week protected their cognition over time compared to nonconsumers, and that effect was even stronger for those who carried a genetic risk factor for Alzheimer's disease.[12] For young folks, fish consumption is associated with higher intelligence scores.[13] And for pregnant women who eat fish, those brain-boosting fats (among other nutrients readily found in fish) lead to stronger brain development for baby.[14]

I've primarily used salmon in my dishes, because it's loaded with healthy fats and is found in most supermarkets. But don't be afraid to change it up and venture beyond salmon. You'll find some other favorites like salted cod (known in Portuguese as bacalao) and sardines in my recipes as well. Here are some of my favorite fish:

Know Your Fish

Tuna	Tuna is a very common genus of fish that encompasses multiple species including yellowfin and albacore. A magnificent and predatory creature, tuna can grow up to 1,500 pounds! It is commonly consumed raw as sashimi, seared, or canned. It's lean (though you can find fatty parts like its prized belly, aka *toro*) and rich in protein. Avoid bigeye tuna, which contains double the mercury levels of other species of tuna.
Salmon (Atlantic)	Atlantic salmon is usually farmed, since wild Atlantic salmon have become endangered. Farmed salmon can contain higher concentrations of certain environmental toxins like PCBs and even flame retardants. It should be stated that farming practices are improving, with nations like Norway leading the charge, and may provide a solution to overfishing. This is a good option if you can't afford or access wild salmon.
Salmon (Sockeye)	Sockeye salmon is the strongest tasting of the salmons, and it's also the leanest (don't worry, it's still a great source of brain-protecting omega-3 fatty acids). Casual salmon fans might opt for the milder choices below, but aficionados see sockeye as a wonderful, well-priced delicacy. What sets sockeye salmon apart is its high concentration of astaxanthin, the pigment responsible for its deep red color. Astaxanthin is a powerful antioxidant, especially beneficial in keeping your eyes, skin, and brain youthful.
Salmon (Coho)	Coho salmon has medium fat content and is a wonderful option for those who enjoy a moderate-flavored salmon. It's typically lighter in color than sockeye and a great mid-priced option for salmon newbies.
Salmon (King)	King salmon is the most prized of the salmons due to its high fat content and rich, buttery taste. It's also the most expensive. It tends to be lighter in color due to the higher fat content, which also means that it contains less astaxanthin, that powerful antioxidant most concentrated in sockeye.
Sardines	An abundant and sustainably caught staple in Mediterranean cuisine, sardines are small, cold-water fish that are rich in omega-3 fatty acids and minerals like selenium, which protect your brain and support thyroid function. Typically bought canned, look for varieties in water, marinara sauce, or extra-virgin olive oil. I enjoy eating them out of the can or in my Better Brain Bowl (page 226).
Cod	Cod is a mild-tasting, low-fat, white-fleshed fish and an excellent source of protein and other nutrients like selenium and vitamin B_{12}. I love to order miso-marinated cod in Japanese restaurants, and cod also makes up the national dish of Portugal (bacalao), which my mother used to make for me. I've provided my own prized family recipe for salted cod on page 184!

EPIC EGGS

There's a reason why everyone in the culinary world has rushed to "put an egg on it"—eggs are delicious! And versatile. They can be made in a dozen different ways, add value to countless recipes, and pack a potent nutritional punch. Unfortunately, entire generations were taught to be fearful of eggs. From the demonization of cholesterol that took hold in the 1970s and '80s to the "your brain on drugs" television campaign, which likened a drug-addled brain to a delicious fried egg, the poor egg's reputation over the years has taken a beating (no pun intended). But, thankfully, the times they are a-changin'.

If there's anything that modern science has taught us, it's that biology is complicated. How chemicals act outside of our bodies is seldom replicated once inside. For instance, the fat we eat doesn't automatically become the fat on our thighs, even though we use the same word for both. The relationship dietary cholesterol has with blood cholesterol (the marker long associated with heart disease) is also not quite linear. You'd be surprised to know that for most people, there's little long-term relationship between the two; you could eat all the cholesterol you want and not see your blood levels budge.[*]

We also now know that an egg may be nature's perfect food, containing a little bit of everything required to grow and maintain a healthy brain. When an embryo is developing, the first structure to coalesce is the nervous system, which includes the brain. For that reason, an egg yolk is like a care package,

---

[*] About 15 percent of the population are hyper-responders to dietary cholesterol, meaning they experience a greater response to dietary cholesterol compared with the rest of the population. However, there are other variables that influence how this affects blood cholesterol, such as the fat composition of one's diet (more on this later). For most people, dietary cholesterol remains a non-issue.

designed by Mother Nature herself, to provide all the nutrition required for a well-functioning brain. It's the non-mammal equivalent of mother's milk, another perfect food *and* one that is also rich in cholesterol.

While cholesterol is distributed throughout your body to every cell and tissue, your brain contains the highest concentration, by far, for its volume: 25 percent of your body's cholesterol is contained in the brain. (The amount of cholesterol in circulation is actually tiny by proportion.) In fact, while your other organs mostly receive cholesterol produced by your liver, your brain actually produces its own. This is how important cholesterol is to the brain, which is to say, it's not by accident that egg yolks—or breast milk—are rich in it.

Before you go chasing down the highest-cholesterol-containing foods you can find, be aware that you don't need to eat any cholesterol for good health— your body makes all that it needs. But foods that contain it, like egg yolks, also tend to contain myriad *other* nutrients that make your brain and body very happy. Eggs in particular are rich in vitamin B_{12}, choline, copper, iodine, and folate, to name a few; each of these are required for healthy brain function. So, eat your eggs and remember, for the majority of people, there is no longer any scientifically valid reason to avoid the yolks.

When it comes to buying eggs, it's always smart to support farming systems that are kinder to the chickens. As food writer Harold McGee put it in his seminal *On Food and Cooking*, chickens in the food industrial complex have been reduced to "biological machines that never see the sun, scratch in the dust, or have more than an inch or two to move." Though their eggs are still healthful (yes, all eggs are healthy!), pasture-raised hens are happier *and* lay eggs that are even better for you, indicated by the thickness of their shells and the orange hue of their carotenoid-filled yolks.

GOOD	BETTER	BEST
Conventional.	Organic and/or omega-3 enriched.	Pasture-raised.

Eggs can be made in so many different ways, it's hard to pick a favorite! While my favorites are poached and over easy, yours may be soft-boiled or sunny-side up. We may both love our eggs scrambled, but you may prefer yours hard scrambled (more well done), while I prefer mine soft scrambled (less cooked, more runny). Eggs are where life begins, so it's fitting that you can prepare them in so many different ways. In a way, their versatility reflects the diversity of life itself.

One of the keys to making good eggs, regardless of how you cook them, is to not *overcook* them. This has a number of important benefits. Overcooking eggs (particularly the whites) releases more hydrogen sulfide (H_2S), the chemical responsible for that unpleasant "egg" smell. It also may damage some of the delicate and otherwise beneficial fats within the yolk, which are sensitive to heat. And overcooked eggs are simply not as delicious.

Take the scramble, for example. When we scramble eggs, we typically try to accomplish the job as quickly as possible. But the best scrambled eggs require a little time and patience. Use low heat and stir your eggs constantly to ensure even cooking. In doing so, you yield creamy eggs with small, evenly distributed curds. Always take the eggs off the heat while slightly underdone (cooked just slightly below your preference), because the eggs will continue to thicken for a few minutes on their own from the residual heat. Sprinkle with a little flake salt and a dash of extra-virgin olive oil, and welcome to scrambled-egg heaven! You may also enjoy my Herb and Avocado Scrambled Eggs (page 107).

Here are some other ways to enjoy your eggs:

Styles of Eggs	
Sunny-side up	Sunny-side up is a method of frying eggs by cracking the eggs directly into a hot pan and not flipping them (which you'd do with over easy), giving the yolk the appearance of the sun. When making sunny-side up eggs, you can cover the pan to hasten cooking (this actually steams the eggs). Make sure the whites are fully cooked, but leave the yolks runny or more custard-like.
Over easy	Over easy is similar to sunny-side up, but you flip the eggs after a few minutes for a nice and even cook on both sides. "Easy" implies the yolk is left runny, whereas "over medium" would be a more well-done egg.
Scrambled	Scrambling is a quick and easy way to both enjoy eggs and combine ingredients that you have around into a substantial meal. Beat the eggs in a mug or bowl. The longer you spend beating, the better. Then place the eggs in a pan (set on low heat) and slowly stir. Remove from the heat just before your desired level of doneness.
Poached	Poaching is one of my favorite ways to cook eggs; they require minimal preparation, no additional fats, and the cleanup is minimal. See page 110 for a primer on how to make perfect poached eggs.
Baked	Baking your eggs is a novel and delicious way to enjoy eggs along with other foods. You'll enjoy baked eggs in my "Cheesy" Baked Eggs with Broccoli (page 105) and Baked Eggs in Sweet Potato "Boats" (page 112).
Boiled	I've never been a fan of boiled eggs, probably because they're so often overcooked (or cooked too fast) and frequently harbor that infamous "egg" smell. However, if you're making them yourself, it's possible to cook them just right and enjoy a healthy and convenient snack. To cook, place eggs in a pot of water and bring to a boil over high heat. Once boiling, immediately turn off the heat and leave the eggs in the water for 10 to 11 minutes to yield hard-boiled eggs (7 to 8 will yield a softer yolk—also delicious!). Remove and place in ice water for a minute, peel, and enjoy.

FAR-OUT FUNGI

From the big portobellos seasoned with salt, pepper, and extra-virgin olive oil that my mother would roast every summer to the button mushrooms I now regularly saute as an accompaniment to meat dishes, I've been a fan of mushrooms my whole life. Each mushroom has a different flavor and texture, so if you're weary of fungi, it's possible that you haven't found the 'shroom for you. (Yes, finding your ideal match is a little like dating!)

Mushrooms comprise a large family of organisms from the fungi kingdom, which is distinct from animals and plants. That's right: mushrooms are not veggies! And each one has a different flavor and texture. For example, portobellos can be seasoned to taste steak-like, whereas the lion's mane mush

room has a flavor and texture similar to fresh crab. Mushrooms can be quick and easy to prepare *if* you know how to prepare them.

One mistake that many people make is rinsing or washing their mushrooms before cooking them. Fresh mushrooms are already up to 95 percent water, and they act like little sponges—some more than others depending on their texture. For example, the gills underneath the portobello mushroom cap or the spines characteristic of the lion's mane love to hold on to water. Any water you add is only going to dilute their flavor *and* cause the mushrooms to boil once you finally get to cooking them, which can lead to sogginess.

To clean your mushies prior to cooking, simply inspect them and brush off any visible dirt or debris. Do the best you can, but don't worry about making them spotless; if eating a little dirt killed people, our species wouldn't have made it very far! Mushrooms also happen to be on the Environmental Working Group's "Clean 15" list, which suggests that we needn't worry about intensive chemical use in their production, though it might be prudent to buy organic if that's available to you.

When cooking, use plenty of fat (ghee, butter, and olive oil are all perfect choices), salt, pepper, and maybe even some garlic, and turn the dial up. Don't be shy! A high temperature will help evaporate any liquid that seeps out, leaving your mushrooms with concentrated flavor and browning them for a delicious texture.

THE FANTASTIC WORLD OF FUNGI

As prized as mushrooms are in the Genius Kitchen, they're equally valued in the laboratory and in the clinic. Certain species of mushrooms that produce a compound called psilocybin are showing immense promise as a treatment for depression, anxiety, post-traumatic stress disorder (PTSD), and even addiction, with major academic research centers like Johns Hopkins Medicine and London's Imperial College leading the charge. What other benefits might mushrooms offer to humanity? Much of the fungi kingdom remains unexplored, equating to a huge opportunity for research. To date, about 5 percent of the 22,000 known mushroom species have proven useful to humans. Of the estimated 140,000 distinct species there are on earth (many of which are unclassified), there may be 6,000 more 'shrooms out there that could provide us benefit. A wonderful documentary to watch on the topic of 'shrooms is *Fantastic Fungi*, which highlights the culinary and medicinal benefits mushrooms have provided to humans thus far.

For such a small, tasty, and low-calorie package, mushrooms also pack a surprising nutritional wallop. They are a source of myriad vitamins, fiber, and antioxidants, and contain compounds called beta-glucans, which may help support a healthy immune system. Perhaps that's why regular consumers of mushrooms have lower risk of both breast cancer and dementia.[15]

These are a few of my favorite 'shrooms:

Super 'Shrooms	
Button	These are the most recognizable and easy-to-find mushrooms. They are versatile with a mild flavor and can be eaten cooked or raw. I enjoy sauteing them in extra-virgin olive oil, with some salt, pepper, and garlic powder for a quick and easy side dish that complements meat dishes nicely.
Portobello	These are the largest mushrooms typically found in your local market. Their caps can be a great keto-friendly bun alternative for your favorite burger and are delicious on their own. Try seasoning with a little salt and pepper and extra-virgin olive oil to make a meaty-tasting side dish.
Porcini	Porcini mushrooms have a strong nutty flavor and are commonly used in Italian dishes. What sets them apart is that they contain high levels of natural glutathione, which is an important antioxidant and detoxifier that our bodies also produce.
Nutritional yeast	Though not technically a mushroom, nutritional yeast is a fungi with a sharp, cheese-like flavor. It's delicious on eggs, in salads, and sprinkled on pastured pork rinds. See it shine in "Pattyless" Jamaican Beef Patty (page 211) and Vegan Carrot Noodle Mac and "Cheese" (page 126).
Lion's mane	A delicious mushroom that has the texture of cooked crab, lion's mane promotes the expression of nerve growth factor, or NGF. NGF is important for the growth and survival of our brain cells. They can be hard to find fresh, but it's worth the hunt.
Cordyceps	These mushrooms may look like little worms, but they are delicious both raw and cooked, with an earthy, sweet flavor. Cordyceps are thought to increase oxygen uptake into cells and therefore can improve energy levels. Some research even suggests they can have a performance-boosting effect.[16]

FANTASTIC FATS AND OMINOUS OILS

Just a short time ago, the word *fat* would send shivers down the spine of any dieter. For years starting in the late 1950s, the public health mantra was that certain kinds of fat were the cause of heart disease, the leading cause of death nationwide. Fast-forward to the 1980s, when the conversation shifted to the fat on our waists, avoiding the nutrient became the overarching ideology, promoted by doctors, our government, the food industry, and even mainstream media.[17] When I was growing up, my neighborhood even had a supermarket dedicated to the movement called F3: Fat Free Foods.

Unfortunately, there never was much evidence that low-fat diets prevent heart disease or promote weight loss, independent of total calories consumed. In other words, dietary fat, in and of itself, can't make you fat *or* sick. Fat is, however, incredibly satiating and satisfying, found in our healthiest foods, and deliberately added to dishes in cuisines around the globe. It unlocks a world of taste from your food, and you simply couldn't thrive without it. As an essential nutrient to good health, we also need fat to absorb fat-soluble vitamins like A, E, D, and K, among other compounds.

WHY ARE WE FAT?

Western culture has allowed a proliferation of foods that are minimally satiating while providing maximal calorie density. Think commercial cereals, breads, cakes, fried foods, pasta dishes, chips, and desserts. These fattening foods, which typify the Standard American Diet, fill you up eventually, but only once you've already overconsumed them. Part of this stems from the fact that such foods tend to be low in fiber, moisture (because moisture decreases shelf life), and protein, which would otherwise slow digestion and assuage your hunger. These foods are also formulated to be pleasurable to taste and easy to chew. When you eat these kinds of "foods"—regardless of whether they're low-fat, low-carb, organic, or any other sort of marketing gimmick—it becomes difficult to pump the brakes. And research shows that when they're *all* we eat, we tend to overindulge to the tune of 500 additional calories per day. That's an entire pound of fat stored every week! By sticking to the kinds of foods I've prepared for you in this cookbook—unprocessed foods that you cook yourself—you'll feel like you're eating more, but you'll actually be eating less, making weight maintenance (or loss, if that's your goal) exponentially easier.

Choosing the right fats, however, is crucial. This is because unlike sugars, which once consumed are either burned off as energy or siloed for later on, fats stick around as the building blocks of your cells and organs, among other things. Take your brain. The fats you eat form the membranes that protect your brain cells. They also influence the susceptibility of those cells to damage. This means that over time, the fats you eat can affect your mood, memory, and resistance to aging and disease.

To understand which fats you should embrace and which you should minimize, a little chemistry lesson is in order (don't worry, it'll be quick). First, each fat or oil in your kitchen is composed of a mixture of fatty acids. At their core, fatty acids are chains of carbon atoms, and how those atoms are connected to one another affects how they behave both in your kitchen and in your body. In saturated fatty acids, those carbons are linked via straight, single chemical bonds, allowing them to pack together very tightly. That's why fats or oils with a lot of saturated fatty acids (butter or beef tallow, for example) are solid at room temperature.

When a fatty acid contains one or more *double* bonds, they are called *unsaturated fatty acids*. Whereas single bonds are straight, double bonds cause a bend in the carbon chain. Unable to pack together as tightly, fats or oils that are predominantly unsaturated are liquid at room temperature (think olive or avocado oil).

While saturated fats are highly stable, unsaturated fats are less so, and they are susceptible to a form of chemical degradation called *oxidation*. Catalyzed by light, heat, and oxygen, oxidation generates compounds called *free radicals*—reactive molecules that can damage living tissue. This process is self-propagating, and over time, a damaged fat can become the "pilot light" that can set entire cellular neighborhoods ablaze, harming nearby proteins, cell membranes, and even your DNA.

Unsaturated fats with only one double bond are safe under normal circumstances—you can cook with them, and they're healthy to ingest. These are called *mono*unsaturated, and they are in fact the primary fatty acids that comprise olive and avocado oil. They're also abundant in grass-fed beef, pork, salmon, and macadamia nuts. But the modern diet has become awash with oils containing fats with *lots* of double bonds. These fats, called *poly*unsaturated fats, are much more prone to oxidation.

To be clear, polyunsaturated fats are not inherently dangerous. If all we ate were foods in their unadulterated state, we'd have little reason to discuss them. But today, people are consuming more polyunsaturated fats than ever before because they're the primary fatty acids found in industrially refined grain and seed oils, like corn, canola, soybean, and grapeseed. These oils are used in innumerable big-brand salad dressings, sauces, spreads, mayos, breads, pasta dishes, fried foods, and even many granola bars. They're simply everywhere.

In whole, natural foods like nuts, seeds, beef, and fish, polyunsaturated fats are perfectly healthy and guarded by molecules that plants and animals develop, like vitamin E, to protect them. (The term "vitamin E" actually represents a team of eight fat-soluble protector molecules so tough they make *The Avengers* look like a modern interpretive dance troupe.) But the production of cooking oils leaves these fats naked and vulnerable. When we overconsume them, we enrich our cell membranes with these highly unstable fatty acids. This contributes to a process called *oxidative stress*, which ages us. And it gets worse.

THE ESSENTIAL FATTY ACIDS

Polyunsaturated fats comprise our essential fatty acids—fats that we need to get from our diets. Two that have become relatively scarce are eicosapentaenoic acid, or EPA, and docosahexaenoic acid, or DHA. These fats are the focus of a huge and ever-growing body of literature highlighting their beneficial effects for a host of health conditions. Studies show that EPA and DHA can help lower fasting triglycerides and prevent thrombosis (blood clotting), which occurs in cardiovascular disease and is a complication in certain viral infections. They can also help soothe inflammation, a defining symptom of our contemporary diets and lifestyles and keystone of modern chronic disease. (EPA and DHA possess this power in part due to their ability to transform into compounds called *resolvins*, which quite literally help to resolve inflammatory mediators in your body.) DHA in particular is one of the body's most important structural building blocks, used throughout, but is especially concentrated in the brain. Research shows that consuming both of these fats on a daily basis (and reducing our consumption of omega-6 fats from grain and seed oils) can improve our chances of avoiding cardiovascular disease, Alzheimer's disease, and certain cancers, and even improve our body composition while we're at it. Many of the recipes in this book were created to contain biologically appropriate amounts of both.

All cooking oils undergo intensive processing to remove their bitter flavors and aromas. The processing of these oils creates a small but significant proportion of compounds called *trans fats*, which are poisonous to your brain and cardiovascular system. The oils are then stored in our kitchens and are heated (or reheated) when we cook with them, becoming vehicles for other toxins, like aldehydes, which are mutagenic (cancer-causing) and damaging to the energy-generating mitochondria in your cells.

When it comes to choosing which fats to use, here's a rule of thumb: if there's an ad on TV for it, you're probably better off avoiding it. These fats are damaging to your health when consumed in high amounts—and, ideally, you should cut them out altogether:

The Ominous Oils	
Soybean oil	Grapeseed oil
Vegetable oil	Safflower oil
Corn oil	Sunflower oil*
Canola oil	

Keep in mind that many of these offenders are hidden in packaged, processed foods (their top sources are salad dressings, spreads, and fried foods). And don't be steered by the red Heart Healthy logo that often accompanies the products that are made with these oils. They are not beneficial to your cardiovascular system—quite the opposite, in fact—and only achieve this designation because they do not contain saturated fat, another misunderstood nutrient.

---

* Typical sunflower oil contains 70 percent polyunsaturated fat, giving it roughly the same chemical composition as grapeseed oil. Some manufacturers have begun using a new, healthier version of this oil called *high-oleic* sunflower oil, which has a lower proportion of polyunsaturated fat (9 percent) and a higher proportion of healthy monounsaturated fat. This gives the oil a fat profile comparable to avocado oil and is thus safe to consume.

OILS AND SMOKE POINT

The processing that commercial cooking oils endure endows them with high smoke points—the temperatures at which the oil will burn and start to smoke. When an oil starts to smoke, it's because particles in the oil (such as the milk solids in butter, or olive particles in unfiltered olive oil) are burning. Though undesirable from a taste standpoint, this has little relationship to the healthfulness or chemical stability of the oil. Ironically, mass-marketed grain and seed oils (canola, corn, soy, and grapeseed, to name a few) boast high smoke points but become toxic at far lower temperatures. In fact, they are often already damaged by the time you ingest them. While a small amount here and there certainly won't kill you, do your health a favor and try to avoid them as much as possible.

Saturated fat possesses both benefits and some potential drawbacks. Thought for decades to "clog" our arteries like grease in a drain, our ever-evolving understanding of the human body dictates that our arteries aren't a mere plumbing system. Instead, we are an elegant chemistry set in which compounds are routinely ingested, broken apart, and transformed. The impact that individual food components have on our physiology often defies logic, and saturated fat is no different.

The most common grievance against saturated fat is that it can raise your cholesterol. But the degree to which it does, and whether that is meaningful to *your* health, isn't always cut-and-dry. How's your midsection looking these days? Do you have the diet of a twelve-year-old boy? Do you smoke? We now know that heart disease risk is multifactorial, influenced by these variables and others. And while too much cholesterol-carrying LDL in your blood can indeed cause problems, it can also be a *sign* of problems. For example, if you're overweight with high LDL, losing weight might cause your LDL to come back down to normal.

No one likes getting a bad report from the doctor. "Am I going to have a heart attack?" we wonder. Our understanding of lipids is ever-evolving. Having high cholesterol numbers should be looked at critically, especially before making any adjustments to your diet or lifestyle, and certainly before taking medicine. There are important questions to ask first. Are you healthy otherwise? If you are overweight, weight loss can help to correct many problems—abnormal cholesterol levels among them. Does your thyroid need a tune-up? Even people with mildly low thyroid levels (subclinical hypothyroidism) can have higher than normal LDL levels, and consuming my thyroid-boosting Kelp Noodles with Brazil Nut Pesto (page 132) may help. Is your fasting blood sugar and insulin, the hormone that keeps your blood sugar within a normal range, healthy? New research points to insulin resistance, the precursor to type 2 diabetes, as being a much stronger predictor of heart disease risk than LDL by itself, challenging the conventional, LDL-centric view.[18] If you're still concerned, two strategies may work. First, decrease your consumption of saturated fat. No need to fear it; just reduce it where possible. Opt for leaner cuts of animal products, and lower your intake of butter, ghee, and coconut oil, prioritizing extra-virgin olive oil and avocado oil instead. Second, increase (slowly!) your fiber intake with dark leafy greens, cruciferous veggies (broccoli, cauliflower, and Brussels sprouts, to name a few), and fresh whole fruit. Fiber is not only filling, it traps excess LDL in the gut, which you then eliminate when you go number two. You may even consider a daily psyllium husk fiber supplement, which can measurably lower your LDL levels.

Thankfully, the dogma that saturated fat is invariably unhealthful has been proven incorrect over the past few decades by large observational studies that show—on average—no negative outcome for either cardiovascular disease or all-cause mortality (death by any cause) after years of consuming a diet high in saturated fats.[19] Even regular consumers of dairy seem to have better health by opting for full-fat dairy, which contains a higher proportion of saturated fat than any other food, compared to low-fat or fat-free dairy (this doesn't mean that low-fat dairy is "bad"—but that fears surrounding saturated fat may be unwarranted).[20]

Certain types of saturated fats may actually *improve* markers of heart disease and cancer risk.[21] Stearic acid, a saturated fat, has demonstrated such abilities, and is found in grass-fed beef and dark chocolate, two ingredients you'll see a lot of in my recipes. Saturated fat also modulates your hormones, powerful messenger chemicals that exert broad influence over your body, including how you look and how you feel. Too little may hurt your ability to keep muscle on or even enjoy sex.

Testosterone is crucial for all genders to build and maintain muscle, have a healthy libido and sexual function, and even enjoy good mental health.[22] In one study, a diet that was low in both total fat and saturated fat reduced total and free testosterone in men; when they reverted back to their original, higher fat diet, their testosterone levels normalized.[23] Similar findings were seen in a study of postmenopausal women; reducing saturated fat also reduced testosterone.[24]

Forget for a second that saturated fat is a normal component of every single healthy, fat-containing food humanity has ever known; it's also the most chemically stable fat available to us in the kitchen. Saturated fats, unlike polyunsaturated fats, contain no double bonds. This means that compared to industrially refined oils, they are much less toxic to cook with. When stocking your Genius Kitchen, fats and oils that are either primarily monounsaturated or saturated are the best choices to go with, *not* polyunsaturated-dominant commercial oils, despite what TV commercials will tell you.

The five cooking fats I recommend as staples are these: extra-virgin olive oil, avocado oil, butter, ghee, and coconut oil. Stock them consistently.

Extra-Virgin Olive Oil

Extra-virgin olive oil is primarily monounsaturated, with about 15 percent saturated fat. This makes it incredibly chemically stable and healthy to cook with, provided you enjoy (or the recipe warrants) its distinctive taste. And while it maintains its healthfulness when used to cook at high temperatures, it's best to use for low to medium temperatures (375 to 400°F) so the oil doesn't burn. It is the only oil that I would consider a dressing oil, meaning you can actually use it as sauce on eggs, vegetables, fish, and even a good steak—a delicious Tuscan tradition that deserves to be adopted widely! See the page 37 for details on how to buy the very best extra-virgin olive oil.

Avocado Oil

Avocado oil is a wonderful source of mostly monounsaturated fat, which, like extra-virgin olive oil, is chemically stable to high temperatures. It differs in two key ways, however: it has a neutral flavor, making it ideal when you want a tasteless oil (such as for baking), and it has a very high smoke point (520°F), meaning it won't burn at high temperatures. Be mindful to purchase from a brand you trust, as many commercially available avocado oils have been found contaminated with other, less healthful oils.

Butter

Butter is a great fat for medium-heat cooking. Though predominantly saturated and chemically very stable, it has a lower smoke point than avocado oil or extra-virgin olive oil (300°F) due to the trace lactose and casein it contains, which burn at high temperatures. As pure fats go, butter is relatively nutrient-dense, containing about 10 percent of your daily needs for vitamin A in a tablespoon serving, in a form that is more easily utilized by your body than plant-based beta-carotene. It also contains vitamin K_2, popularized by holistic dentist Weston A. Price. Vitamin K_2 is an otherwise hard-to-come-by nutrient that helps deposit calcium in places you want it (bones and teeth) and keeps it out of places you don't (arteries and kidneys). (See page 42 for more on vitamin K_2 and the "calcium paradox.") Opt for grass-fed varieties, which have a higher nutrient density.

Ghee

Ghee is clarified butter; it begins as butter and is gently boiled so that the milk solids rise to the top. They then get skimmed off, leaving behind a pure, tasty, and versatile fat. Because of its purity, ghee has a much higher smoke point (450°F) than butter, and it's even well tolerated by many people who are dairy sensitive. It's a delicious staple in Indian cooking and can make a wonderful addition to a whole host of cuisines. I particularly enjoy using ghee to sear liver (see Bangin' Liver, page 192) and to cook steaks in (see page 209).

Coconut Oil

Coconut oil has seen tremendous waves of interest over the years. Some have demonized the oil due to its high concentration of saturated fat, while others have touted it as a "miracle cure" for everything from tooth decay to Alzheimer's disease. The truth is somewhere in the middle; it does contain medium-chain triglycerides, which can support brain energy metabolism, a potential benefit for those with neurological conditions, but it's probably wise not to go overboard with this tropical fat as it is low in essential nutrients and may adversely affect blood lipids in some. For cooking, it's a great fat to use for low to medium heat (smoke point 350°F). With its strong flavor, it's best used in tropical-tasting dishes or desserts (try it in my Chocolate Blueberry Clusters, page 270).

WHAT ABOUT CULINARY OILS LIKE SESAME AND WALNUT OIL?

Some of my recipes call for sesame oil or walnut oil, which are made simply by pressing the seeds or nuts. Unlike industrial seed oils, culinary oils like sesame and walnut oil do not undergo intensive processing, since the intent is to keep the seed or nut's unmistakable and irreplaceable flavor. As a result, they maintain high levels of vitamin E and other antioxidants, making them perfect to dress your foods with, or even cook with at low to medium temperatures. You can see sesame oil shine in my Japanese-Style Beef, Sweet Potato, Miso, and Mushroom Stew (page 223) and walnut oil in my delicious Watercress, Avocado, and Grilled Fruit Salad with Toasted Almond Dressing (page 120).

EXTRA-VIRGIN OLIVE OIL

Humans have been pressing olives to harvest their delicious and health-promoting oil for thousands of years, and it continues to be a staple in many parts of the world, particularly in the Mediterranean, where rates of cardiovascular disease and Alzheimer's disease are low. Complex and buttery, the juice of the olive fruit presents a flavor profile that is as sophisticated as it is versatile, making it perfect to cook, and dress, a wide variety of foods. By the end of this section, you'll know how to pick the very best of the best.

The healing properties of extra-virgin olive oil come in part from its fat profile; it is rich in chemically stable monounsaturated fat. The particular type found in olive oil is called *oleic acid*. Little did early taxonomists know, oleic acid is the most common fatty acid found in nature; grass-fed beef, wild salmon, avocados and avocado oil, macadamia nuts, and human breast milk are all rich in oleic acid, which derives its name from *olea*, the Greek word for olive. It is the type of fat most beneficial to consume liberally.

Unlike other oils that contain oleic acid, however, extra-virgin olive oil contains an additional superpower: a compound called *oleocanthal*. This phyto- (plant-derived) nutrient is responsible for the peppery flavor that a high-quality extra-virgin olive oil is known for. Imparting more than just a fiery kick, oleocanthal has been found to dampen inflammation similarly to low-dose ibuprofen, a synthetic nonsteroidal anti-inflammatory drug (NSAID).[25] This gives extra-virgin olive oil anticancer and heart-protective effects, without any of the negative side effects associated with NSAIDs.

I love to get to know my oils as I buy them, and an oil's aroma can tell you a lot about the quality of the oil. Whenever you purchase a new oil, the first thing you should do is crack it open and place a tablespoon or so in a shot glass (a wineglass is also fine). With one hand, cradle the glass from underneath, and then place your other hand on top, covering the rim of the glass. Gently swirl the oil around, almost as if it were whiskey in a snifter. The heat of your hand will activate some of the fragrant, volatile compounds in the oil. After ten seconds, put the glass up to your nose and remove your hand. A fresh oil should smell grassy and clean, not waxy like crayons or stale walnuts.

Now it's time for the taste test. A true olive oil connoisseur may practice *stripaggio*, the art of drawing air and the oil simultaneously through the lips and teeth to spray the back of the mouth and throat. Aside from impressing onlookers, this allows one to simultaneously taste the oil and assess the pungency of oleocanthal, that prized, peppery, anti-inflammatory compound. Whatever you do, ensure that it hits the back of your throat on its way down. A good extra-virgin olive oil can be so spicy it may make you cough; the number of coughs is actually used to signify the quality of an oil! A two- or three-cough oil and you know you've found a keeper.

When it comes to purchasing olive oil, there are a few important things to know. Perhaps the most obvious but often overlooked is to make sure your olive oil is "extra-virgin." This means that the oil has undergone no additional steps beyond the pressing of the olive fruit and filtering of the oil. "Light" olive oils not only lack flavor but are devoid of the antioxidants found in extra-virgin oil and are processed similarly to grain and seed oils, like corn or soybean. Ensure that your oil is not a blend containing other, cheaper and less healthful oils, like canola. And skip unfiltered oils—this just means that tiny olive remnants are left in, but these can accelerate degradation of the oil and lower its smoke point.

Next, packaging matters. While canola, corn, and soybean oil (sometimes labeled "vegetable" oil) are usually sold in clear plastic which can leach potentially harmful petroleum-based chemicals like bisphenol A (BPA) into the oil, olive oil producers have reverence for their product and want to protect it. As a result, you'll often find it sold in glass. Choose dark bottles to keep your oil protected from harm as you enjoy its flavor and health benefits.

One more consideration to make when choosing an extra-virgin olive oil: size. Unlike a fine wine that may appreciate over time, the quality of extra-virgin olive oil only degrades. It's a perishable food, after all! Not only do you want to look for a harvest date on your bottle (the more recent the harvest the better), you'll want to purchase your oil in sizes small enough to use within a month or two, knowing that once you open up the bottle and expose the oil to oxygen, the degradation process advances. The perfect size for you will depend on how much you use—start with one liter and see how far it takes you!

Now that you know how to pick the best oils, I encourage you to regularly try different brands. A true connoisseur will have their favorites, but you'll never develop your palate if you stick to the same old oil. Buy often, and spend the most you can comfortably afford. A quality olive oil is a prized possession, a way to delight your taste buds, enhance your health, and impress friends and loved ones (you can even keep a bottle as a centerpiece on your dining-room table!).

Here are some of my favorite uses for extra-virgin olive oil:

- **To saute.** Olive oil is a perfect fat to use at temperatures typically used to saute (low to medium heat). Start with 1 tablespoon of oil (or more depending on the recipe) and spread it around. It's a wonderful starter for sauteing garlic, onions, mushrooms, and more. Keep in mind that without any food in the pan, any setting on the stovetop—even low—will cause the oil to get hotter and hotter and eventually burn. Make sure to put your ingredients in the pan before the oil burns.

- **As a sauce.** I love to drizzle extra-virgin olive oil on all types of dishes, from eggs to steak. It's a perfect sauce, as it enhances the flavor of whatever you put it on but does not overpower. Keep in mind that olive oil still is calorie dense, healthful as it is. Generally, 1 or 2 tablespoons per day max is plenty (that's 120 or 240 calories, respectively).

- **To roast.** I love to toss veggies in extra-virgin olive oil and then roast them. If I'm cooking at 400°F or lower, it's the perfect fat to complement cauliflower, Brussels sprouts, squash, and more. Make sure to add plenty of salt, and maybe pepper and garlic, and you'll never consider veggies dull again.

- **As a spread.** I occasionally buy gluten-free sourdough bread or breads made with nutrient-dense flours like almond flour (sometimes called "Paleo" breads). While some grass-fed butter is a good topping, if you've never tried dipping your bread into extra-virgin olive oil, you've been missing out. Try placing the oil in a small plate and adding a splash of balsamic vinegar and maybe a sprinkle of flake salt. Divine!

- **In dressings.** Extra-virgin olive oil is the primary oil I use in salad dressings. It lends such a strong foundation to which you can add lemon juice, your favorite vinegar (see page 57 for some options), salt, pepper, garlic . . . there are no rules! You can even use it by itself, topped with some flake salt—perfect for sliced tomatoes, an avocado, or even some grapefruit slices! Writing this is making me hungry.

ONE MORE WAY FAT IS ESSENTIAL FOR HEALTH

When you observe the range of beautiful colors represented by fresh produce, you may be happy to know that they're not just there to appease your eye. Many of those pigments are part of a group of compounds called *carotenoids*, and each confer powerful health benefits. For instance, sweet potatoes and carrots are both rich in beta-carotene, giving them their orange hue. Beta-carotene is converted in your body to vitamin A, an antioxidant that is essential for healthy eyes, bones and teeth, and immune function. Tomatoes contain lycopene, a red carotenoid, which protects your cardiovascular system and, fellas, your prostate. Then there's lutein and zeaxanthin, found in foods with yellow tones like kale, avocado, peas, and fresh corn. These compounds help keep your brain sharp as you age and can even help prevent age-related macular degeneration. The catch? All of these compounds need to be consumed with fat to be absorbed and utilized in your body. No fat, and these compounds flow right through you—a huge missed opportunity! So ditch the fat-free salad dressings and drizzle extra-virgin olive oil liberally on your salads (1 to 2 tablespoons is fine) and coat your veggies in it prior to roasting. (If you're consuming veggies with another fat source—a handful of nuts, for example—there is no need to add more fat.) Now keep calm and carrot on!

DAMN GOOD DAIRY

From dairy begins all life—it is the only substance designed by nature to be food. Created by mother, milk delivers protein, carbohydrates, and fat for baby's nourishment, growth, and development. It also sends over specific messages about the state of the world. For example, stressed mothers pass the hormone cortisol through their milk, perhaps readying her offspring for a turbulent environment. It's so elemental to our being that the word *mammal*—the taxonomic class including cows, bats, whales, and us—literally means "of the breast."

Fast-forward to adulthood, and dairy is now a ubiquitous food group that includes milk, yogurt, and cheese. While breast milk may be the perfect food for an infant, questions remain about dairy's role in adulthood. Some may think it's odd that humans are the only species that continues to drink milk past weaning, and they'd be right—but let's be real. We're also the only species that flies in airplanes and thumbs away our time on smartphones.

Dairy is perhaps most famous for the calcium it contains, and it is indeed a great source (calcium is the most abundant mineral in the body and important for growth). But contrary to what you'd assume from its publicity over the years, dairy does not hold a monopoly on the mineral. Sesame seeds, dark leafy greens like kale, almonds, and sardines (with the bones in them) are all great sources of calcium.

THE CALCIUM PARADOX

Are weak and brittle, calcium-depleted bones the result of not consuming enough calcium or the result of other bone-building variables being out of place? We now know that vitamin D is important for calcium absorption and that resistance training is crucial—crucial!—for building strong bones. But we've also begun to recognize the role of the little-known nutrient menaquinone, or vitamin K_2, for bone health. Found in the fermented soy dish natto, the fat of grass-fed beef, butter, ghee, dark-meat chicken, and pastured egg yolks, vitamin K_2 is instrumental in activating bone-fortifying proteins like osteocalcin. This helps deposit the calcium that you may already have in your body where you actually want it to go—i.e., your bones and your teeth. Vitamin K_2 also helps activate other proteins, like matrix Gla protein, which helps escort calcium out of soft tissues, like your kidneys and arteries. Similar to the uneven distribution of our planet's economic or environmental resources, it's becoming clear that weak bones may be a symptom of poor calcium distribution, rather than an overall deficit of calcium—and vitamin K_2 may offer a redistribution solution! Enjoy my Coconut Braised Chicken Thighs with Mustard Greens and Cauliflower (page 205) or my Bangin' Liver (page 192), recipes with an especially high vitamin K_2 content.

There are some potential downsides to consuming dairy as an adult. Many people don't tolerate it past infancy, when many of us (about 75 percent, globally) stop producing the enzyme lactase, which breaks down a sugar found in milk called *lactose*. Without lactase, lactose ends up in the large intestine, where it gets fermented by the microscopic party animals in your large intes-

tine, aka your gut microbiome. This leads to unpleasant symptoms like bloating, gas, and diarrhea (gee, thanks guys).*

For those who do tolerate dairy, the question remains: should you consume it? Large observational studies have shown us that people who consume more dairy seem to have better health. Most surprising is the finding that people who choose *full-fat* dairy are better off than those who opt for low-fat or fat-free varieties.[26] Still, these are correlational findings. It's also possible that low-fat dairy products are more likely to be loaded with added sugar—something to watch out for if you choose to get down with dairy.

There is one perk to dairy that is undeniable: when it comes to protein, dairy is spectacular, possessing among the highest biological value protein to be found. The two major proteins in milk are called *casein* and *whey*. Some believe the former to be a driver of inflammation, though recent meta-analyses suggest that dairy is not pro-inflammatory, at least for the general population.[27] Still, some are sensitive to casein and may feel better excluding it from their diets.

There is another context in which dairy may be useful: cooking. Butter and ghee (introduced on page 36) are wonderful cooking fats, and you'll find ghee used in a small handful of my recipes. (Butter is of course a wonderful food on its own, but the scientific evidence favors extra-virgin olive oil as the more healthful dressing of choice for your food.) Both butter and ghee are very low in casein and lactose, with ghee being the purest and therefore most suitable for those with dairy sensitivities.

---

* If you're lactose intolerant, you can try supplementing with the enzyme lactase to enjoy occasional dairy. In fact, many dairy producers make lactose-free versions of yogurt, sour cream, and even milk by simply adding the enzyme lactase to the product.

All things considered, I advise moderation for dairy consumption (my personal dairy consumption ranges from no dairy at all to one or two servings per day). Some great high-protein options include Greek yogurt, cottage cheese, and whey protein shakes. When buying full-fat items, try to look for grass-fed options (second best would be pasture-raised). If buying a low-fat or fat-free product (see below for why this could be a smart option), there's no need to opt for grass-fed—just make sure you choose plain varieties with no added sugar.

Here are some dairy-based foods to keep your kitchen stocked with.

Dairy We Love	
Butter	Butter is a rich source of vitamin A and vitamin K_2. It is made from cream that has been churned. The churning process chemically alters the cream to make the dense and delicious fat we call butter, but this also makes it more prone to raising cholesterol. It comes in salted and unsalted varieties. Ideally purchase butter from 100% grass-fed cows.
Ghee	Ghee is butter that has been clarified. It has a distinct taste and is a perfect cooking fat to make liver in or to use in desserts. As with butter, use in moderation.
Greek yogurt	I love Greek yogurt. Full-fat varieties contain healthful fat-soluble vitamins, but fat-free is a very-low-calorie, high-protein food. It ultimately comes down to your goals. If you're trying to lose weight, opt for fat-free because it will be lower calorie. Always reach for plain, unsweetened varieties, and sweeten with stevia or whole fruit.
Whey protein	I always keep whey protein on hand. A simple protein shake can be a rewarding and satiating snack and can also increase your daily protein intake, which helps support muscle growth and maintenance. Whey protein can also boost the production of glutathione, which is your body's master detoxifier and antioxidant.[28] I prefer vanilla-flavored protein powders because chocolate varieties may harbor higher levels of lead, arsenic, and cadmium, according to a 2010 *Consumer Reports* study. If dairy sensitive, choose whey protein *isolate*.
Hard cheeses	Hard cheeses are nutrient-rich and lower in lactose than soft cheeses. They also tend to contain higher levels of vitamin K_2 and other unique dairy-based nutrients. I don't have a hard-and-fast rule to cheese consumption. . . . If you enjoy and tolerate it, feel free to eat it in moderation!

NEXT-LEVEL NUTS AND NUT MILKS

Nuts are an incredible food and have been a reliable part of the human diet since the beginning of history. They're nutrient-packed, delicious on their own either raw or roasted, and can be used to make any number of products once ground: nut milks, nut butters, and grain-free breading and baking products, to name a few. More than just tasty, the health benefits of nuts are significant. Observational studies routinely point to a risk-lowering effect of nut consumption on cardiovascular, neurodegenerative, and respiratory diseases, cancer, and other ailments.[29]

Each nut has a unique nutrient profile. For example, almonds are a major source of magnesium and vitamin E, important youth-promoting micronutrients that most of us (50 and 90 percent, respectively) do not consume enough of. Brazil nuts are a powerhouse of selenium, an important antioxidant and mediator of healthy thyroid function (one single Brazil nut provides more than your entire daily need for selenium!). Pistachios are abundant in carotenoids, pigments that give them their unique yellowish-green hue, which

also happen to protect your eyes and brain from aging. And macadamias are the nuts with the highest proportion of heart-healthy monounsaturated fat.

It doesn't take many nuts to give you a health benefit—just over half an ounce, less than a handful, per day (or 4 to 5 ounces per week).[30] That's good news, since the one potential downside of nuts is their energy density. Next to pure oil, nuts are among the most calorie-dense whole foods that exist, owing to their high fat content. Take almonds, which are 54 percent fat—one single handful can contain upward of 400 calories! The relative ease with which they can be overconsumed (we no longer have to forage or shell them for ourselves) makes eating nuts by the fistful treacherous ground for the weight conscious. And then there's nut butter.

Have you ever eaten almond butter out of the jar? No? Well, let me tell you, it's hard to stop once you get that train rolling. Nut butters are tasty but can also easily be overeaten, so be careful. I keep my house stocked with peanut butter (peanuts are not actually nuts but legumes, but for the sake of simplicity I'll include them here) and almond butter and occasionally use them as toppings. Though they are good for you, I'm always cognizant of serving size. Try to stick to 1 to 2 tablespoons, max—and always ensure your nut butters are free of added sugars and oils.

Last, there's nut milk. When soaked nuts are ground up and strained, they leave behind a milky solution that is a perfect replacement for dairy milk. Nut milk is easy enough to make at home, but thankfully there are many premade nut-milk options on the market now. When choosing a nut milk, ensure that yours is unsweetened, free of added oils, like canola or grapeseed, and contain no synthetic emulsifiers (carboxymethylcellulose and polysorbate-80 are the two most suspect). Animal studies have suggested an inflammatory effect of these chemicals on the gut.[31]

Depending on your body goals, you might also glance at the calorie content of your nut milk. Most of the "OG" commercial nut milks contain just a few nuts, lots of water, and thickening gums. As an alternative, artisanal products that use many more nuts per bottle are now on the market. Though delicious and less processed, some of them can be excessively high in calories, *even* if unsweetened. If you're watching your weight, reach for the lower-calorie versions (up to 40 calories per cup), which "get the job done" and allow you to put those calorie savings toward actual food.

WHAT'S THE DEAL WITH GUMS?

Food-grade gums are found in myriad processed foods and are usually used as a thickening or emulsifying agent. You'll often find them in man-made foods like nondairy milks and ice creams where manufacturers need to keep fat dispersed in an aqueous (water-based) solution. There are a ton of gums: gellan gum, xanthan gum, guar gum, and locust bean gum, to name a few. These gums generally are concentrated in tiny amounts but can nonetheless cause gastrointestinal distress in some people. I recommend limiting foods that contain them (partly because they tend to be ultra-processed, hard-to-resist junk foods), but they're generally harmless if a little bit sneaks in here and there, and you don't notice a sensitivity to them.

How should you incorporate nuts? Use them in recipes that naturally limit how many of them you can eat (try my Brazil Nut Pesto on page 132 for starters; it's a winner!). Since they are so fatty, they go wonderfully on lean fish and chicken dishes, in salads, and in desserts. Coating fish with pistachio nuts or throwing some slivered almonds into a salad may irreversibly uplift your perspective on those two dishes. Nuts also make great gluten-free flour, which can be used to coat chicken tenders or create any number of baked goods. When purchasing nuts, ensure raw or *dry* roasted, as "regular" roasted nuts typically are coated in unhealthy oils. These are a few of my favorites:

Nuts about Nuts

Almonds	Almonds are incredibly versatile; they're the nut most commonly used in alternative flours and nut milks, and they're easy to snack on. They're rich in vitamin E, magnesium, fiber, and calcium, and their skins have been identified as a potent fuel source for beneficial gut bacteria.
Brazil nuts	Brazil nuts are very calorie dense, containing almost 70 percent oil. In fact, just two nuts can contain the caloric equivalent of one egg. That said, they're tasty and a wonderful source of selenium, with one nut containing roughly double the recommended daily intake. Stick to no more than 3 to 4 per day.
Coconut	Coconut isn't a nut but a fruit and is sweet and unctuous and lends flavor to many cuisines around the globe. It's lower in essential nutrients and higher in saturated fat than any nut, so consume it in moderation, especially if you have genes that might predispose you to heart disease. Coconut oil is rich in medium-chain triglycerides, which may be helpful to those with neurological conditions.
Pistachios	Pistachios are delicious, and their health benefits are unique. They contain more lutein and zeaxanthin than any other nut (helpful to protect your eyes and brain) and are rich in potassium and vitamin B_6.
Macadamia nuts	Whereas most nuts are high in polyunsaturated fats, macadamia nuts are abundant in monounsaturated fat, a more chemically stable unsaturated fat. They are also very calorie dense and tend to be the most expensive of all nuts.
Peanuts	Peanuts are not actually nuts but legumes. That said, they're usually found in the nut-butter aisle. They're higher in protein than other nuts but would not be considered a high-protein food because they're still mostly fat by calories (almost 50 percent).

HEROIC HERBS AND SPICES

Herbs and spices may not add much volume to your dishes, but don't let their small footprint fool you. They are the on-ramp to food that makes a statement, adding big and bold flavors that will delight your taste buds and impress family and guests. Spicing and seasoning your food not only makes bland ingredients mouthwatering, it also adds negligible calories. The same can't be said for sauces, which are usually calorie bombs, loaded with fat or sugar or both.

Small but mighty, herbs and spices have been used for centuries by various cultures for their medicinal value and add a potent health boost to any meal in which they're used.

Whether to improve digestion, reduce pain, or even help heal wounds, many staples of the contemporary spice rack have long been revered by ancient medicine. Thankfully, science is beginning to catch up, showing us that these culinary must-haves are in fact teeming with unique bioactive compounds that possess broad, health-promoting effects.

While there are hundreds of spices in use around the world, only a few have been the focus of rigorous research, and much of the experimental evidence is performed in petri dishes or on critters. Of the handful that find their way to human, clinical trials—garlic, ginger, cayenne, and cinnamon are a few of the most well studied—there does seem to be broad, anti-inflammatory, anticancer, neuroprotective, and blood-sugar-lowering effects. This may explain why people who consume spicy food daily have a lower risk of early death by 14 percent![32]

Time will tell whether spices prove to be a panacea to human health, but in the meantime, here are some of my favorites to keep your kitchen stocked with (I've used many of these in the recipes to come).

Spices to Stock

Black pepper	Cayenne pepper
Cinnamon	Coriander
Cumin	Garlic
Ginger	Mustard seeds
Paprika	Turmeric

Herbs to Hoard

Basil	Cilantro
Parsley	Oregano
Rosemary	Mint
Thyme	Sage

Using herbs in your food is another great way to add flavor and an additional health kick. Dried herbs work best at the beginning of the cooking process or for longer cooking times, whereas fresh herbs work better to add in toward the end of cooking. The reason for this is that fresh herbs have volatile compounds in them that are destroyed by cooking, and depending on the herb, it may be those distinctive compounds that we want. That's also why we prefer some herbs dried and some fresh. For example, fresh parsley and basil are nonnegotiable in certain dishes, but oregano works just fine as a dried spice in most dishes that call for it.

Whether you're a seasoned pro or herbally timid, I implore you to free your mind and push your flavor boundaries to new heights with herbs and spices.

STOKED ON SALT

Whether it's meat, fish, eggs, veggies, or even salads, salt is the magic that takes single ingredients and transforms them into *food*. And yet it is another misunderstood kitchen staple due to widespread fears linking the sodium

it contains to hypertension, or high blood pressure. It's time for a reality check: sodium is an essential nutrient, important for regulating blood volume and nerve and muscle function among other things. So widely utilized in the body, sodium is actually classified as a *macromineral*. This means we need to consume a relatively large amount of it every day for good health.

The majority of us already consume enough salt, but it doesn't come from our cooking *or* our saltshakers; it comes from processed, packaged foods ranging from canned veggies and soups to frozen dinners and fast food. Volatile compounds in fresh food lose their potency over time, so manufacturers add in tons of salt to compensate. Plus, it's a very effective preservative. Today, Westerners overconsume these kinds of foods, comprising 60 percent of the calories we ingest every single day. And it isn't even obviously salty foods that are the top source; the number one source of dietary sodium in the Western diet is bread and rolls.

Sodium is also overused in restaurants. Because restaurants rely on shelf-stable goods, their dishes are often a one-two punch: sodium-enriched foods with salty coatings, crusts, and sauces, and that's not even accounting for the breadbasket. According to the Centers for Disease Control, packaged, processed foods and restaurant foods combined are the source of 71 percent of the sodium the average person consumes. Only 11 percent of the salt we ingest comes from home cooking.* The salt that you add to your food at home is *not* the problem.

Cutting out salt also won't do much to lower your blood pressure if you're like the majority of the population who are sodium insensitive, but doing so can have other undesirable effects as the body attempts to compensate for the loss of sodium. A very-salt-restricted diet can raise levels of the hormone insulin and may make you temporarily insulin resistant.[33] It might also raise your cholesterol and the amount of fat in your blood (triglycerides).[34] As the science around diet and heart disease continues to evolve, the path to better blood pressure clearly favors minimizing consumption of packaged foods, losing weight if overweight, exercising regularly, sleeping well, and managing your stress over just not salting your food.

---

* We ingest 6 percent of our daily sodium from home food preparation, and 5 percent is added at the table, totaling 11 percent. If you're wondering where the remaining sodium comes from, the answer is simple: food. Many of our healthiest foods, such as beets, spinach, celery, and beef, are natural sources of sodium. Yet one more reason why the demonization of this mineral is misguided.

For most of us, it's not the salt per se that drives high blood pressure; it's the overconsumption of packaged foods. These foods don't just lead to sodium overconsumption; they make our bodies actually hold on to it as well. Here's how: foods high in refined sugar and flour (i.e., your typical packaged food) lead to chronic elevations of the hormone insulin, which causes your body to retain sodium, raising blood pressure in return. Cutting these types of food out (and cooking at home more frequently) will likely cause your sodium levels to plummet and your blood pressure to normalize.

Some of us (about 26 percent) are more sodium sensitive than others (that statistic is naturally higher for the people with hypertension). This means that their blood pressure rises by 5 percent or more when they switch from a low-sodium diet to a high-sodium diet. Being overweight (particularly if you carry lots of visceral fat, or fat around the midsection) can also drive insulin levels up, which is why weight loss often helps to lower blood pressure. And finally, increasing potassium intake can help "cancel out" sodium's impact on blood pressure. Some great, low-sugar sources of potassium include beef, salmon, avocados, dark leafy greens, squash, and potatoes (no, fries don't count)— many of these and others are used in the recipes that follow.

With your fears around salt allayed, remember that salting is the number one way to make healthy, albeit bland, foods taste good. I couldn't imagine eating broccoli, or cauliflower, or root vegetables without salt, and these foods are incredibly good for you! Try feeding Brussels sprouts without salt to a child, and they'll probably wince (I know I would have as a tot). Even the most well-marbled rib eye is a missed opportunity without salt. Always use salt in your cooking, and season with salt to taste.[*]

When it comes to stocking your kitchen, you should have three types of salt: fine, coarse (or kosher), and flake, a finishing salt. "But Max, why on

---

[*] The only exception to this would be for infants, who are not efficient sodium excreters. They get all the sodium they need (and can handle) via breast milk and what is naturally found in food.

Earth do I need three types of salt if they're all the same thing?" Simple: It's one of the easiest and most inexpensive ways to elevate your cooking to a restaurant-quality level of sophistication. Below is what you need to know.

Types of Salt	
Fine salt	Fine salt is what most people think of as table salt. It's the smallest grain, ideal for cooking, recipes, and the saltshaker. It can be precisely measured and dissolves quickly. Keep some in your kitchen and on your table, though you can keep it off your table if you use flake salt (see below), which is what I do.
Coarse salt	Coarse salt, aka kosher salt, is a thicker grain and is perfect for jobs that require a more rugged salt but still require an even consistency. Salting meat prior to cooking is the most classic use-case scenario (see page 212). This is usually kept in the kitchen and not used at the table.
Flake salt	Flake salt is a finishing salt. You do not cook with it; you simply sprinkle your food with it just before serving. The grains are large and salty and add an irresistible crunch to anything they're sprinkled on. I keep a dish with flake salt on my dining-room table at all times, and even ask for it in restaurants (every good chef uses it, and you can often find it in restaurants if you just ask). It is sublime on all meats, fish, eggs, veggies—literally everything!

In a perfect world, our salt would be harvested from a pristine sea, but today, that's a pipe dream. Many commercially available sea salts have been found to contain tiny plastic particles—the unfortunate reality of mining our increasingly polluted oceans. And industrially refined salts are not much better; though some are a good source of iodine, they can be packed with anti-caking agents and sugar (yes, most iodized salts contain dextrose, a sugar, to stabilize the iodine). I personally like pink salts, but they can be hard to find in all three of the grain sizes I mentioned above, and I often use sea salt as a result. In the end, the best salts to use are the ones you can find and afford.

VERY AWESOME VINEGAR

Maybe I'm weird (okay, I'm weird), but vinegar is one of my favorite flavors. Whether it's balsamic vinegar, apple cider vinegar, or rice vinegar, chances are if it has vinegar in it, I'm into it. Originally discovered thousands of years ago by early winemakers (the word *vinegar* actually derives from the French word *viniagre*, or "sour wine"), it is now a staple in kitchens around the world, including the Genius Kitchen.

Widely used as a preservative and pickling ingredient, vinegar also makes a delicious and tangy addition to any number of recipes. It's used to poach eggs, makes the foundation for a wonderful marinade, and adds definition to sauces; even ketchup wouldn't be ketchup without the subtle tang of vinegar. People drink it, too. The acidic kick of kombucha, a fermented tea that has become a $480-million-per-year business in the US alone, is owed to vinegar. See the following recipe for an easy way to dip your toe into the "drinking vinegar" pool.

The Vinny Gambini

This is a refreshing tonic to lower inflammation, balance blood sugar, provide brain-protecting metabolites, and maybe even help control your appetite. You can enjoy this hot or cold (for a hot drink, just heat the water—but not the vinegar—and omit the ice). Drink on its own or sip with a meal.

Serves 1

1 tablespoon apple cider vinegar
½ teaspoon ground or fresh ginger
¼ teaspoon ground cinnamon

2 drops liquid stevia or monk fruit sweetener
¾ cup water
Ice

Combine the vinegar, ginger, cinnamon, stevia, and water in a glass, stir to dissolve the flavorings, and serve over ice.

No matter which type you choose, vinegar possesses potent health benefits thanks to its primary ingredient, acetic acid. Its most well-documented benefit is to your blood sugar and insulin levels. When added to your meals, vinegar can lower their glycemic impact and reduce the amount of insulin that your pancreas secretes.[35] This makes it a potentially useful tool to help fight type 2 diabetes, which affects almost half of the people in the US (including those with prediabetes), and insulin resistance, which likely affects many more.

Here are some of my favorite vinegars and their uses.

Types of Vinegar

Distilled white vinegar	Though made from fermented grain, white vinegar contains no calories or carbohydrates and is considered gluten-free no matter which grain it comes from. I don't often use it for cooking, though it is my go-to for poaching eggs (see page 110 for a simple recipe). It can be added to laundry for an extra sparkle and when diluted makes a wonderful safe home cleaning product.
Apple cider vinegar	Made from the fermentation of apple cider, ACV, as it's popularly referred to, contains beneficial enzymes and polyphenols, which can support gut bacteria. It may also have an appetite-suppressing effect, making it easier to eat less if weight loss is your goal.[36] It's awesome in salad dressings, on roasted veggies, and in beverages. Purchase raw varieties that contain the "mother" (rich in probiotic bacteria) for maximum health benefits.
Balsamic vinegar	Made from fermented grapes, balsamic vinegar is a must in any kitchen. Its slightly sweet flavor comes from a small amount of naturally occurring grape sugar. It also contains trace amounts of resveratrol, a compound studied for its longevity-promoting potential, and 3,3-dimethyl-1-butanol (DMB), which promotes a healthy gut. I love it in salad dressings and to occasionally drizzle on steaks with a hefty shot of extra-virgin olive oil.
Red wine vinegar	Red wine vinegar is a staple in Mediterranean-style cooking. It has a mildly fruity flavor and is at home in any Spanish, Greek, or Italian dish. I particularly like to use it in conjunction with heartier flavors. You'll find it in my Baked Eggs in Sweet Potato "Boats" (page 112) and Avocado, Fennel, Pomegranate, and Winter Citrus Salad (page 156).
Rice vinegar	Rice vinegar is primarily used in Japanese cooking. I like to keep some in my cupboard to add to takeout sushi rice for a little extra kick, and you can also use it to make Japanese sunomono dressing: combine ¼ cup rice vinegar with ½ teaspoon tamari sauce or coconut aminos, 2 tablespoons sugar-free sweetener of choice, and ½ teaspoon salt (delicious to toss sliced cucumbers with!).

RUN TO THE WATER

Water is the stuff of life! Our planet is abundant with it, our food is rich in it, we cook with it, we drink it (we're supposed to, anyway), healthy digestion relies on it, and we are made of it. Yes, about 60 percent of your body weight is water. Each organ—your brain (73 percent water), your muscles and kidneys (79 percent water), your skin (64 percent water), and your heart (71 percent water)—relies on water to function properly. In this section I'm going to share with you how to stay properly hydrated so that you can perform optimally, digest your food effortlessly, and feel your best.

When you think of water, what comes to mind? The simple and ubiquitous molecule H_2O, comprising two hydrogen atoms and one oxygen? Or perhaps the wet stuff that comes out of your sink's faucet? If you're like most people, you think of the latter, which is anything *but* pure water. Tap water is, like most water found in nature, a veritable soup of compounds! For instance, it regularly contains dietary copper, calcium, and magnesium. In fact, water may provide up to 20 percent of the required daily intake of calcium and magnesium, and for copper, some tap waters contain a full day's worth.[37]

Clearly, tap water can contain very important constituents, but it has also been found to be polluted all over the US, containing trace levels of heavy metals like arsenic, endocrine- (hormone) disrupting chemicals like fluoride, pesticides, and even pharmaceuticals. These chemicals tend to appear below their legal limits, but the science on how daily ingestion of low doses can affect your health is far from settled. Many such chemicals, like bisphenol A (BPA), a plastic-related compound that mimics the sex hormone estrogen, are not routinely monitored for. And they may have unforeseen effects at low doses—a looming question in the field of toxicology.[38]

To be sure, hydration trumps dehydration, and one of the greatest achievements of the eighteenth century is running water—make use of it. But if you'd like to add a layer of protection between you and the xenobiotic (foreign) chemicals that have been identified in tap water, there are a few inexpensive options available to you.

The lowest barrier to entry is carbon filtering, which is inexpensive, effective at removing pollutants, and easily accessible. This typically comes in the form of pitcher filters or filters that are built into your refrigerator. These systems use carbon granules to trap various contaminants, including some heavy metals, chlorine, and pesticides. Many do not remove fluoride (a potential endocrine disruptor) and may allow some solids (including metals) to remain. Still, being the lowest cost water purification device, it's better than nothing, and some are more thorough than others. Always look for lab reports, which reputable brands make available on their websites.

If you have a bigger budget for water purification, opting for a reverse osmosis (RO) purifier may make sense. RO purifiers are the most thorough, as they clean your water of nearly all metals, pesticides, pharmaceuticals, chlorine, bacteria, viruses, and fluoride. They achieve this by forcing water molecules through a semipermeable membrane that leaves everything else behind. The downside is that these systems also remove important minerals. This may not be ideal for most folks, since the Standard American Diet is already deficient in many essential nutrients. However, if you consume a nutrient-rich diet and have the budget and counter space for an RO purifier, it may be the way to go. You can also remineralize your water after it's been purified with seawater-derived drops, which can be found online.

When it comes to hydration, plain water isn't the only option—good news for the 7 percent of adults who don't consume *any* water daily. Beverages like tea and sparkling water are also hydrating. Just be sure that your drinks don't have any sugar—drinking your calories is not a good habit, since caloric beverages provide no satiating effect and yet can significantly eat into your daily calorie budget. Also, be sure to minimize drinking out of plastic or cans that might have a BPA inner lining, since each can leach harmful hormone-disrupting chemicals into your drink.

Get Fizzy Wit' It Sugar-Free Soda

Who doesn't like soda ("pop" for you midwesterners)? The problem is, sugar-sweetened beverages like soda underlie the obesity epidemic along with all of the associated harms that come with being overweight. This is my take on a zero-calorie soda that is delicious and hydrating as well as inexpensive, quick, and easy to make.

Serves 1

1 can (BPA-free) orange-flavored sparkling water

4 to 5 drops vanilla-flavored liquid stevia

Crack open the can of sparkling water and drop in the stevia drops through the mouth hole. Gently swirl, then drink. Feel free to try different flavors of sparkling water—black cherry and coconut are also great!

We also get a significant amount of water from our food. In fact, meat is more than half water, and fresh fruits and vegetables are up to 95 percent water. The latter can be incredibly hydrating and perfect on those hot or physically demanding days when pounding the water just doesn't seem appetizing. Stews and broths are other fantastic options. One other benefit of "eating" your water: foods rich in water are satiating, provide lots of minerals, and yet have very low calorie density, meaning you can eat more without dipping significantly into your calorie budget! Magnifico.

Here are some of my favorite water-rich foods:

Food	Water content
Cucumber	95%
Watermelon	92%
Bell pepper	92%
Broth	92%
Strawberries	91%
Cantaloupe	90%
Citrus	88%
Yogurt	88%
Beef (cooked)	~60%
Chicken (cooked)	~60%

SUGAR, STARCHES, AND ALTERNATIVE SWEETENERS

It goes by many names: cane sugar, coconut sugar, high-fructose corn syrup, honey, and molasses, to name a few. Regardless of what clothing it wears, sugar—in the functional sense—is sugar. There might be trace, hard-to-quantify beneficial compounds in foods like maple syrup and honey, but ultimately, it's all the same.

There's been a lot of confusion as of late over where sugar fits into an optimal diet. Some argue that any amount is toxic and should be avoided, and yet some of our healthiest foods—fresh berries, for example—contain natural sugars. Others say that as long as you don't overeat it, even "added" sugar in the form of high-fructose corn syrup can be fine. The truth, as where many questions in health land, falls somewhere in the middle: sugar isn't inherently bad or good—its role in your health depends on context.

When looking at the kinds of food sugar is normally found in, there are two primary categories: fresh produce and processed foods. Fresh produce, including whole fruits and veggies, normally contains sugar in varying quantities. The sugar in these foods isn't much of a concern because fresh produce contains myriad beneficial compounds, and the foods themselves are self-limiting due to their fiber and water content. There are only so many apples you can eat before you start to feel full, or sick (imagine eating ten apples—much harder to do than eating a box of cookies, and the sugar content is roughly the same).

WHEN EVEN FRUIT SHOULD BE MODERATED

Many people have trouble metabolizing sugar because they have diabetes. Diabetes is a problem involving the hormone insulin whereby the body either can't produce enough (type 1) or it has become insensitive to it (type 2). For those with diabetes, dietary sugar causes blood sugar to stay elevated for too long, leading to damage of the blood vessels and organs. For this population, which today is at least one in every two adults, not only should added sugar be avoided, but even dense sources of natural sugar—sweet grapes and tropical fruits, for example—can present a problem. You'll be glad to know that the dishes and sweet snacks I've created don't involve any added sugars, and I've specifically chosen low-sugar fruits for those that incorporate fruit. I've also crafted my desserts using sugar-free sweeteners, which taste sweet but provide no calories and (according to the current research) do not influence hormonal fluctuations that can perpetuate hunger, which sugar can do. These sweeteners, such as monk fruit and erythritol (a sugar alcohol), are naturally derived, minimally processed, and much better tasting than the noncaloric sweeteners of the past.

Processed foods—including the foods we process ourselves, via cooking—are a different story. Added sugar and syrups are essentially empty calories; they provide no nutritional value, and yet eat into our daily calorie budget, the number of calories you can consume in a day before gaining weight.* Were you a bodybuilder with a massive calorie budget (having more muscle

---
* Calories are like money: you ingest a certain amount each day, and you burn (spend) another. You generally want calories in/calories out to be balanced so that you don't store fat. Eating sugar is like spending your money on stuff you don't need, leaving less money for the essentials (i.e., less room in your calorie budget for the foods that nourish you).

on your body means a higher resting energy rate) or a professional athlete, some sugar here and there would be no big deal. But most of us (and I'm including myself here) are not bodybuilders.

Today, your average human spends their day inactive, especially relative to their ancestors, and inhabits a world where intense physical activity is all but absent. As a result, our energy requirements are modest. This allows added sugar to easily displace the calories from more nutritious food in our diets. With the average person now consuming sixty-six pounds of added sugar per year, it's no wonder 90 percent of us are now nutrient deficient. This handicaps our bodies' ability to fight aging and disease, which means that eating less sugar is an important step on the path to staying youthful and healthy.

Sugar is also a major cause of our ever-expanding waistlines. Unlike protein and fat, our bodies have no specific requirement for dietary sugar, and so our appetite for it can be a bottomless pit. We rarely tire of eating it, and we have no hard-wired sensors to tell us we've had enough. On the other hand, protein, fat, and fibrous carbohydrates from whole, minimally processed foods each play a role in making you feel satisfied. This is why minimizing foods with added sugar (and especially sugar-sweetened drinks) is pivotal to losing weight. Unless you do, you can easily consume your entire daily calorie budget from sugar and *still* be hungry for real food.

Here are some of the many names that sugar goes by. Cut back on foods with these ingredients and your health and waistline will thank you:

The Many Names of Sugar

Beet sugar	Brown sugar
Cane sugar	Date sugar
Dextrin	Glucose syrup solids
Maltodextrin	Evaporated cane juice
Fruit juice concentrate	Corn syrup
Invert sugar	Malt syrup
Honey	Maple syrup
Molasses	Brown rice syrup

There's one more downside to sugar: what it does to your teeth. Teeth are crucial to the survival of an animal; a toothless animal is quickly a dead one. Today we've come to accept widespread tooth decay as "normal." Sugar directly feeds bacteria in our mouths that cause cavities and periodontal disease, whether it fits into your personal calorie budget or not. And while sugary drinks may actually be less harmful from this vantage point—liquids don't linger in the mouth long enough for bacteria to latch on—sugary foods like desserts, candies, and grain flour–based products may be the most offensive as a result of their natural retention between your teeth.[39]

Refined Grain Products to Avoid

Bagels	Crackers
Doughnuts	Muffins
Granola bars	Chips
Gravies	Waffles
Pizza	Pancakes
Breads	Cakes
Oatmeal	Wraps
Whole wheat products	Multigrain products

Although modern grain flours don't look or taste like sugar, they're quickly broken down into sugar molecules, a process that begins upon chewing. Then, they stick to your teeth—a perfect storm for tooth decay, especially when accompanied by sugar. According to a recent review of food types and their cariogenicity (cavity-causing potential), researchers wrote: "Starches can increase the cariogenic properties of sugars if they are consumed at the same time." So while any starch can promote tooth decay, sugar-loaded grain products like cakes, cookies, crackers, granola bars, commercial breads, and breakfast cereals are among the worst offenders.

The recipes in this book don't use any refined grain flours or sugars (except for the small amount sometimes found in dark chocolate). Occasionally, I'll use small amounts of cassava flour or tapioca starch, which is derived from the cassava plant. Cassava is a tuber, and a proportion of its starch is resistant to digestion. This type of resistant starch is plentiful in uncooked roots and tubers, unripe fruits, and in some nuts and seeds, and is less likely to feed unhealthy bacteria in the mouth, acting instead like a fiber with numerous health benefits.

WHY YOU SHOULD COOL YOUR STARCHES

One of the most surprising food "hacks" I've shared over the years is that a cooked-and-cooled potato is better for you than a freshly cooked potato. Why? Because the cooling of a cooked potato (or any starch) increases the amount of resistant starch it contains. Resistant starch is resistant (as its name implies) to digestion, acting like a form of fiber instead. This helps reduce the caloric and blood sugar impact of the starch, can help improve blood lipids like cholesterol, and also promotes healthy gut bacteria. Resistant starch is naturally present to varying degrees in raw potatoes, unripe bananas, papaya, and mango (my personal favorite), cashews, cassava, and other starch-dense plants, but cooking (or ripening) these plants causes the starches to break down into sugars that are easily digestible once consumed. When cooled, however, some of the starch "retrogrades," reverting back to its original, indigestible form. This makes foods like rice, potatoes, oats, and even pasta better for you once cooled to room temperature (or colder). And the good news? You can still reheat them to enjoy and maintain their newly formed resistant starch. Plus, the more times you cool and reheat, the more resistant starch you generate. Love your carbs? Don't cool your jets; cool your starches instead!

Thankfully there are a slew of sweetener options that don't add calories or feed disease-causing bacteria in our mouths for us to use as well. While the first noncaloric sweeteners were synthetic and, frankly, didn't taste very good, today we have many naturally derived sweeteners that not only do not add any calories or sugar to our food but taste great as well. (Some, like erythritol and xylitol, in fact directly support oral health by helping prevent cavity formation.) In my recipes I've offered numerous sweet treats using my favorite noncaloric sweeteners.

Here are the types I've chosen to use:

Types of Sweeteners

Monk fruit	Monk fruit has been used medicinally for centuries in Chinese medicine, where it is known as *luo han guo*. There are some alleged benefits to using monk fruit, including healthier blood sugar control. Monk fruit can have an aftertaste, and for this reason is often combined with erythritol, a sugar alcohol (more on this below). Throughout this book, where a recipe calls for "monk fruit sweetener," I am using such a blend.
Sugar alcohols	Sugar alcohols are naturally found in many foods and, despite their name, are not sugars, nor will they give you a buzz. The only two sugar alcohols I recommend are xylitol and erythritol, which are very well tolerated by the gut, even in high amounts (many other sugar alcohols can cause uncomfortable GI symptoms). Both bake and taste like sugar, though they can impart a "cooling" effect on the tongue if used in high amounts.
Stevia	Stevia comes from the South American *Stevia rebaudiana* plant. Safe to use and with no caloric contribution, stevia is great to use in small amounts, but it can have an aftertaste. For this reason, you'll often find stevia mixed with other sweeteners, like erythritol or allulose (below). I keep vanilla-flavored stevia drops (available at most health food stores) in my kitchen to add to plain Greek yogurt or flavored sparkling water to make a quick and easy zero-calorie soda (see page 60 for a simple recipe).
Allulose	Allulose is a newer noncaloric sweetener derived from nature. It is very well tolerated (it won't upset your stomach) and tastes and feels just like sugar without adding calories. Though more research is needed, early studies indicate a potentially beneficial blood sugar–lowering effect.

That sums up all you need to know to stock your Genius Kitchen with food. But buying good food that fills you up and makes you feel great is only half the battle. Up next, an odyssey into the methods of cooking, eating, and storing.

12 SHOPPING TIPS FOR A GENIUS KITCHEN

1. **Shop the perimeter.**

   Most supermarkets are laid out in the same way. The perimeters of most supermarkets are where you'll find the fresh produce, meat, seafood, eggs, and dairy. These are the foods that make up the basis of what I cook for myself, and of what I've prepared for you. No shade on aisles; they also contain important ingredients like spices, dried herbs, and vinegars. But for the bulk of your shopping, the perimeter is where the good stuff is.

2. **Quality over quantity . . . to a point.**

   I'm a stickler for quality. I generally prefer to buy higher-quality ingredients than a larger number of discretionary items. For example, I'll usually buy wild salmon over farmed salmon, or grass-finished beef over grain-finished beef. However, if you can't find or afford these "best-case" options, please know that grain-finished beef and farmed fish are still better options for dinner than boxed mac and cheese. Everyone's budget and access are different; your goal should not be perfection but to do the best you can.

3. **Buy whole poultry and cheaper cuts.**

   You'll get the most bang for your buck by buying whole poultry and cheaper cuts of meat. With poultry, you can cut up the bird and freeze what you don't immediately use. With beef, you can slow cook, or simply cut against the grain to tenderize a tougher cut of meat. And don't forget organ meats (discussed on page 15)—these are the supermarket's best kept secret for nutrient density!

4. **Don't overspend on organic.**

   Shopping organic reduces your exposure to synthetic pesticides, but it's not a panacea, particularly if the product in question is just ultra-processed junk with an organic certification slapped onto it. If you can easily afford organic, go for it. If you'd like to be more deliberate with your dollar, a good rule of thumb is: if you eat the entire fruit or the skin or peel (apples or berries, for instance), consider organic. If organic is totally off-limits, make sure to rinse your produce well.

5. **Become fluent in "nutritionese."**

   Knowing how to read a label is one of the most important things you can do for your health. Ingredient lists are written in descending order of concentration. This means that the first few ingredients are the most abundant in a given food. Be mindful of serving size, the "added sugar" indication (ideally, that number is zero or close to it), and the number of servings in a container. If your food has sugar in the first, second, or third slot (revisit page 64 for a list of some of sugar's possible alter egos), skip it, as that means it has a ton of empty calories.

6. **Don't be duped by marketing.**

   You'd be surprised to know just how many health claims (either explicit or implied) your average shopper is exposed to. "Organic!" "Low-fat!" "High-protein!" "Heart-healthy!" "Keto!" The truth is, this is marketing for what are often hard-to-resist junk foods. You won't see health claims on broccoli, avocados, wild salmon, or grass-fed beef, which are among the healthiest foods in the supermarket and adhere to most of these claims anyway! In general, stay away from health claims.

7. **The fewer the ingredients, the better.**

   While the number of ingredients a product claims has no bearing on whether it's good for you, the quantity of ingredients does correlate to the degree of processing that the "food" has undergone, and as a rule, ultra-processed foods are to be minimized. Real, quality foods do not have extensive ingredient lists; they *are* the ingredients. If the product in question contains more than eight ingredients, in most cases you might consider a different (less-processed) variety instead.

8. **Not all processed foods are unhealthy.**

   I know I like to harp on about the dangers of food processing, but the reality is, food processing isn't all bad. When you grill, fry, broil, bake, steam, or blend your food, you are in fact processing it! While it's certainly smart to be wary of the food industry and its many newfangled creations, some processed foods are great: avocado oil–based mayos, sugar-free ketchup, nutritional yeast, pre-sliced turkey breast (nitrite-free), and dark chocolate are just a few of my favorites that I couldn't live without.

9. **Make friends with frozen.**
Buying frozen fruits and veggies is a great way to stretch a dollar. Since they are usually frozen at the point of picking, their nutrients are locked in. And because they don't need to be rushed to your supermarket the way fresh produce does (usually by air), frozen produce is more affordable. Grass-fed beef and wild salmon are also popping up in the freezer section at major supermarkets. Don't fear frozen!

10. **Shop online.**
Today there are many Internet and subscription-based companies that deliver very-high-quality meat products and even better-for-you processed products like grain-free breads through the mail and are often less expensive than what you'd get in a supermarket (and, of course, you can't beat the convenience). New options pop up all the time, so visit my resources website at http://maxl.ug/GKresources for the latest suggested vendors and exclusive discounts.

11. **Make a list and stick to it.**
We all know the dangers of heading to the supermarket and veering off the path of our intent. This is why it can be helpful (and healthful) to make shopping lists before you go. Jot down a list of your nonnegotiables: eggs, unsweetened almond milk, dark chocolate, a pint of blueberries, and a bag of frozen wild salmon are examples of staples that are usually on my list. Make it a goal to think through your list and stick to it.

12. **If it's in your shopping cart, it's in your stomach.**
Think of your shopping cart as your extended stomach. When considering that bag of greasy chips, throw out the idea that you'll be able to eat them in moderation. (While some people are able to stick to just a handful, many won't.) Visualize the whole bag, in your body, wreaking havoc on your hunger and health, and move on.

CHAPTER 2

CREATING GENIUS HABITS

Everything should be made as simple as possible, but no simpler.

—ALBERT EINSTEIN

In the previous chapter, you discovered how food can interface with your body, from meat, to fish, to herbs and spices, and even water. But a Genius Kitchen is about so much more than the food you stock it with. Ensuring that you have the right tools and rituals at your disposal guarantees that your food choices work for you, not against you, making your body as happy as your mouth.

In this chapter I will cover a few of the underlying principles that guide cooking in your Genius Kitchen, from must-have utensils to methods of food preparation. Then you will discover how to fully enjoy the fruits of your labor by learning ways to optimize your digestion so that you feel great during and *after* your meal.

DON'T BE A TOOL; USE THE *RIGHT* TOOLS

Stocking a Genius Kitchen takes more than just food; it requires tools. When it comes to selecting the best cookware and utensils, we want to strike a balance between efficacy, practicality, and low toxicity. What's the point of cooking healthy food if you're using cookware that leaches toxins (hormone-disrupting chemicals, for example) into your food? And what good are non-toxic supplies if they don't *work*, or cost an arm and a leg to buy?

When stocking your kitchen, keep it simple. A pan, basic set of pots, knives, a sheet pan, measuring utensils (cups, ½ cups, ⅓ cups, ¼ cups, tablespoons, teaspoons, and fractions thereof), tongs, spatula, spoon and slotted spoon, and cutting board are the bare essentials. Because these are the items that come in direct contact with your food, the materials they're made with matter.

Utensils for a Genius Kitchen	
Spoon	Instant-read thermometer
Slotted spoon	Measuring cups
Spatula	Measuring spoons
Tongs	Cast-iron pan
Cutting board	Knife set

As a general rule, a Genius Kitchen aims to reduce plastic. Not only is plastic a major source of detritus filling our landfills and oceans, it is made using petroleum-based compounds, which are able to migrate into food under certain circumstances. One such chemical is bisphenol A, or BPA. BPA's hormone-disrupting qualities have been known since the 1930s, when it seemed destined to become an estrogen replacement drug. When it was discovered that BPA could be used to make plastics, it infiltrated human life and now appears in everything from our furniture to our cookware.

Even BPA-free items can still harbor hormone-scrambling compounds, such as chemically similar variants bisphenol S and bisphenol F. It has become a game of chemical Whac-A-Mole, but there is no reason to believe that replacement products are any safer than BPA. The best way to avoid them is to minimize the use of plastic, especially when that plastic is meant to have prolonged contact with hot, acidic, fatty, or salty food.*

Wood and silicone are great options for durable and nontoxic utensils. Silicone has much higher durability than plastic, low risk of chemical reactivity, and is free of petroleum-based compounds like BPA, BPS, and the like. Both wood and silicone are safe to use for high-temperature cooking and won't scratch delicate cookware.

For cookware, look for stainless steel, ceramic, or cast iron. A cast-iron pan is a chef's must-have item, in part because it withstands high temperatures and distributes heat so evenly. A well-seasoned pan also has nonstick characteristics, assuming you've gotten it hot enough prior to putting food such as eggs into it (more on this on page 79). Stainless steel and ceramic are other great options with high heat tolerance and low chemical reactivity.

WHAT'S THE DEAL WITH NONSTICK?

Nonstick pans were devised in the 1960s and are traditionally made with chemicals called per- and polyfluoroalkyl substances, or PFAS. These chemicals were hailed for their cooking properties but have since become a pressing environmental and health concern. Because of their durability, these chemicals never break down, and are now found nearly everywhere—in our drinking

---

* Plastic can be safe; just be mindful not to overheat it or allow hot food to stay in contact with it for longer than a few seconds. Consider replacing old, worn-looking plastic utensils with new ones.

water, in our food, in animals, and in us. Exposure to PFAS chemicals can lead to health problems, including thyroid dysfunction, low libido, immune system issues, cancer, and developmental problems. While the most well studied of them have been phased out, it's unclear whether replacements are safe, especially since real-world use of items made with them often involves exposure to unsafe temperatures and scratching from metal utensils and other cookware. If you choose to use a nonstick pan, look for one made without PFAS chemicals, and treat them delicately; do not overheat (medium heat is fine and makes great eggs), use only wood or silicone utensils on their surface, hand wash, and do not stack metal pans inside them, which could lead to scratching. Inspect your pan regularly and replace if there are any signs of wear.

When it comes to storing wet, hot, salty, or acidic foods, opt for glass instead of plastic. Glass does not leach any chemicals into your food. What do you do if your glass containers have plastic tops? If they don't come in contact with your food, you've got ninety-nine problems, but the lid ain't one. If you already have plastic containers, rather than throwing them away, you can repurpose them for dry goods, provided they're not stored in a hot or damp environment.

As you can see, "it depends" is the answer to many questions in science—and that includes whether to use paper or foil. For general use, such as for baking a rack of Perfect Ribs (page 202) or laying down some Chocolate Blueberry Clusters (page 270), unbleached parchment paper is the winner. Just make sure not to burn the paper; temperatures below 420°F are fine. For higher temperatures, wet ingredients, and use near an open flame, aluminum foil is obviously more practical and safer; just know that under certain circumstances, foil can allow aluminum to migrate into your food.

Aluminum is not naturally found in us and serves no purpose in the body, and though the health effects of aluminum are unclear, the precautionary principle dictates the less we ingest, the better. Unfortunately, the aluminum in foil (as well as in aluminum baking pans) is not inert; it can contaminate our food, which we then ingest. The primary catalyst for the leaching of aluminum is contact with salty and/or acidic marinades, a fact illustrated by an alarming German study.

THE PRECAUTIONARY PRINCIPLE

The concept of the precautionary principle originated in a 1992 United Nations consensus on environmental protection that asserted, "Where there are threats of serious or irreversible damage, lack of full scientific certainty shall not be used as a reason for postponing cost-effective measures to prevent environmental degradation." In other words, absence of evidence is not a valid excuse for inaction. Applied to health, the less time a food or product has been in circulation, the greater the burden of proof it should bear before it is considered safe. Too often that responsibility falls onto consumers (as opposed to manufacturers) with consequences to follow. While occasional exposure to aluminum is probably harmless, there is simply a lack of evidence to say with certainty that long-term, low-level chronic exposure is absolutely safe. And as the precautionary principle dictates, the less we expose ourselves, the better.

The researchers found that citric acid, the primary acid in lemon and other citrus, could easily absorb aluminum—150 milligrams of it—and that heating it to 320°F for two hours increased the aluminum levels even higher, by twofold to threefold.[40] The authors estimated that a daily intake of 8 ounces of fish marinated in lemon juice in an aluminum dish could lead to consuming 187 percent of the European Food Safety Authority's Tolerable Weekly Intake (TWI) for an adult and 871 percent for a child. While aluminum foil is not considered dangerous, think twice before marinating or cooking acidic foods in it.

One material with no ability to leach into your food is wood, and the final must-have utensil for the Genius Kitchen is a wood cutting board. Not only is a good cutting board important to preserve your knives when slicing and chopping away, it makes a wonderful platter to serve on, especially for meats. After pan-searing a steak, I love to place it on my cutting board to rest. Then I sprinkle on some flake salt and bring the entire board out to my guests. Visually, it's wonderful, and the wood keeps the steak warmer than a cold ceramic plate would.

SAVE MONEY NOW BY REDUCING FOOD WASTE

Did you know that the average American family of four throws out $1,600 in wasted produce every year? Most of it is sent to landfills, where it undergoes fermentation by bacteria, contributing 7 percent of the world's greenhouse gas emissions. But with a few tricks up your sleeve, you can save money and the planet by reducing food waste. Here are some of my favorite tips:

- Store sliced carrots and celery in water. Celery will last up to two weeks longer and carrots up to one month longer.

- Have herbs you haven't used in time? Prevent them from wilting by freezing them in an ice cube tray submerged in extra-virgin olive oil.

- Life threw you too many lemons? Juice them and freeze the juice in ice cube trays.

- Not in a rush to use your potatoes? Keep them away from onions and store them with an apple to prevent sprouting.

- Avocados brown too darn quickly, but refrigerating them and brushing some lemon juice on the exposed flesh if already cut open will help slow browning.

- Up to your ears in onions? Pickle them. Slice and place in a jar with vinegar and a pinch of sugar-free sweetener (I use a monk fruit and erythritol blend).

- Berries aplenty and no time to eat them? Wash them with diluted vinegar to extend their life up to two weeks.

- Kale is already a hard sell for most, without being limp. Breathe new life into your kale by cutting its stems and placing it in water in the fridge for a few hours.

THE INVINCIBLE IRON PAN

Yeah, Iron Man is cool, but have you ever used an iron *pan?* Growing up, we had a cast-iron pan passed down at least one generation. By the time it was in my mother's possession, it had become a prized family heirloom . . . one that was also used to cook incredible dishes. Cooking with and maintaining a cast-iron pan can be tricky, but it's worth the effort. Not only is iron chemically unreactive under normal cooking circumstances, but it distributes heat steadily and evenly. It can also be placed inside an oven, which is useful for finishing thick steaks. Plus, it's cheap, and as my grandma Hilda used to say, a penny saved is a penny earned!

Unlike other pans, cast-iron pans form a seal on the surface with repeated cooking of fatty foods that makes them nonstick (yes, even for eggs). This occurs because over time, the oils oxidize and form an inert coating over the iron that protects it. This seal is called a "season." The oils quickest to do this are the ones that contain mostly polyunsaturated fats (revisit page 27 for a primer on fats). This is why corn, soy, and particularly flaxseed oil, which is the *most* easily oxidized due to its abundance of uber-delicate omega-3 fatty acids, are the most commonly used to season a pan.

Ultimately, any oil will create this seal over time, so cooking on and caring for your pan appropriately will do the job without the need to buy these oils. To season (or re-season) your pan, simply rub the entire surface with the liquid oil of your choice—enough to leave a thin, even layer. Then bake your pan in the oven for thirty minutes at 450°F (you can repeat this for a stronger seal). Once cooled, your pan should be left with a shiny, black, waterproof gloss.

Creating a seasoning is easy enough; maintaining it is the hard part. Keep your cast-iron pan out of the dishwasher and use a stainless-steel scrubber or scouring sponge to clean it. Simply hold your pan under a running faucet and lightly scrub (without soap), ensuring a clean, residue-free surface. Alternatively, you can scrub the pan with coarse salt as an abrasive to lift stubborn bits. You can do this with or without water. In either case, dry with a towel or simply heat the pan on your stovetop. This continues to season the pan with each use. Cast iron–pan users are passionate and opinionated about their care, so if you are going to go down this route, it could be fun to check online for the various theories!

When it comes to cooking with a cast-iron pan, ensure that your pan is hot *before* placing ingredients in it that would otherwise stick (you can also use some oil, but keep in mind that fatty foods like steaks and burgers have plenty of their own fat to use). With these tips, your cast-iron pan should last a long time and be part of the creation of many delicious dishes. Who knows—maybe you'll even pass one down to your kin!

YOU'RE PROBABLY OVERCOOKING FOOD

Food should be delicious and nourishing; it shouldn't make you sick. Food safety is an important part of that equation. Foodborne pathogens—whether bacteria or parasites—can indeed make you quite ill. But is raw or rare meat or fish inherently dangerous, or are these food taboos yet another unusual artifact of the modern food supply? In this section I hope to allay the fears that cause you to overcook your meat and fish and to point out that under-cooking *and* overcooking both have risks worth heeding.

Industrial food production is an astonishing feat of ingenuity, and yet like many things in life, it is a double-edged sword. It helps many of us have access to fresh meat and produce no matter our geography or time of year. But it can also be an assembly line of imprecision, where a single unfriendly microbe can infect huge swaths of product. Food production is now highly mechanized, and this includes the preparing of animals and fish for human consumption. This is precisely why it's very risky business to eat undercooked meat or fish today.

On the other hand, there's nothing implicitly dangerous about fresh, raw meat. Properly sourced and prepared fish can make wonderful sashimi, and Mediterranean chefs will often prepare pesce crudo ("raw fish") from fresh fish with a little extra-virgin olive oil and sea salt (you can easily try this at home with sashimi-grade fish). And though it is most common to do this with fish, it's not the only type of meat that can be eaten raw. While enjoying a weekend at the farm of one of my favorite producers of grass-fed, grass-finished beef, I enjoyed raw beef for the first time.

Different cultures have their own takes on raw beef; you may have tried steak tartare or beef carpaccio. My first experience of it was in a dish called carne cruda ("raw beef"), the Italian version of steak tartare, prepared by my friend and renowned chef Anya Fernald. She used freshly sourced, chopped lean beef (lean meat is ideal when consuming raw beef; raw fat feels waxy in the mouth) and then drizzled it with extra-virgin olive oil, threw on a generous pinch of flake salt, and a hefty squeeze of lemon. Though I was reluctant at first, the dish was unexpectedly divine. I couldn't help myself from going back for seconds and thirds.

This may surprise you, but raw chicken is not implicitly dangerous, either; in fact, chicken sashimi (aka torisashi) is commonplace in Japan. But industrial production makes such "delicacies" impossible. Chickens have small bodies, and crowded pens and hurried evisceration (the reality for most chickens, sadly) spells very, very high risk of fecal contamination, where pathogens like *salmonella* can end up in places they shouldn't. This is why you should never consume undercooked chicken or other poultry unless you're absolutely certain it's safe.[*]

Of course, raw meat and fish aren't for everyone, and there are few things more mouthwatering than a perfectly seared steak or salmon fillet, or the skin of a well-cooked bird. Still, you should be able to enjoy your meat without overcooking it. Take a nice, juicy steak, for example: while the USDA

---

[*] Note from my lawyers: do *not* try to eat undercooked chicken at home.

recommends cooking all beef to a medium-well internal temperature of 145°F, steak aficionados know that the juiciest and most flavorful steaks are cooked to medium-rare with a temperature of between 130 and 135°F.

There are important health benefits to not overcooking your food as well. Slow cooking methods that use lower heat tend to preserve the health-promoting qualities of the fats in your food (recall that foods like salmon and grass-fed beef are rich in polyunsaturated fats, which are healthful in small amounts but can be denatured at high temperatures). Meat and fish also contain powerful antioxidants like vitamin E that degrade with high heat cooking.

Cooking your food can also generate harmful compounds if not done carefully. Polycyclic aromatic hydrocarbons (PAHs) are one such example. They are generated when organic material of any kind (even plants) is cooked to the point at which it burns. Heterocyclic amines (HCAs) and advanced glycation end products (AGEs) are also formed on the surface of well-done meat. Though dangerous to cells in petri dishes, none of these compounds should be worrisome in the context of an antioxidant-rich diet that also includes veggies—but precautionary principle dictates that the fewer of them there are around, the better.

The key to cooking your food safely is to follow proper food prep hygiene, and that *begins* with sourcing. Between factory farms and mega-supermarkets, our separation from food production means that, on average, people simply have no idea how their meat is handled prior to purchasing it. Find a butcher, learn his or her name, and discover where the food you're eating comes from—it can make purchasing higher-quality meat an incredibly gratifying experience, and it promotes mindfulness of the animal that has been sacrificed for your benefit. And a skilled butcher always knows the freshest and safest cuts of meat, the ones that won't need to be overcooked to be enjoyed.

Always inspect the food that you buy. The most potent way of gauging freshness is smell. Get to know the scent of fresh food and food past its prime. Fresh meat should never smell. Over time, bacteria can gather on the surface of meat and release malodorous gases like hydrogen sulfide (you may detect a whiff of egg smell in decaying meat) or can give off a hint of ammonia. (Bacteria can also make fish smell fishy, which you read about on page 16.) Next, feel it; does it feel slimy or sticky? That's a bacterial biofilm and means it's going bad. Check the sell-by and use-by dates. Meat can usually be consumed for a few days after the sell-by date, but don't consume meat past its use-by date.

After you've found your meat or fish, always make sure your surfaces and utensils are clean when cooking. Soap and warm water are effective for your hands and dishes, and you can make a simple cleaning solution to wipe down countertops using undiluted white vinegar which is effective at killing *E. coli* and *salmonella*, two of the most common food-borne pathogens (unfortunately vinegar is not effective against viruses). And when preparing meat, make sure that *its* surfaces are thoroughly heated, since surfaces are what attract bacteria—and chances are, in the supermarket, it already has.

By following these steps, you should be able to enjoy your food cooked to perfection *without* having to burn it into oblivion. Now, *mangia*!

The ancient Greek physician Hippocrates, who is often considered the father of modern medicine, is believed to have said, "All disease begins in the gut." Your gut, which includes your digestive tract, allows you to break down and assimilate the nutrients in your food, and digestive problems (low stomach acid, for example) can impair this process, leading to malnutrition. It is also an important site of detoxification, where unsavory compounds such as heavy metals, including lead, cadmium, and myriad other xenobiotics, or foreign substances, are trapped and purged.

Your gut also serves as the dumping ground for lipids and hormones that your liver deems excessive. For example, "used" LDL particles and hormones such as estrogen could cause problems if allowed to build in your body to excess. (Having too many LDL particles in your blood is thought to be a predictive risk factor for heart disease, and too much estrogen can increase risk for breast cancer.) Astonishingly, your gut helps balance both, where they hitch a ride to the toilet—*if* your digestive system is running smoothly.

As you can see, your gut is a living, dynamic system; it's not simply a biological garbage compactor, though many treat it as such. It responds to various cues—including time of day and even your mood—which can affect how well it functions. It also caters to the trillions of bacteria and viruses that live within it—known collectively as the gut microbiome—an exciting area of science. (If you'd like a thorough and hype-free understanding of how the microbiome affects your health, I encourage you to read my book *Genius Foods*.)

In light of this, the next section will focus on how to optimize your digestion at the point of consumption: mealtime. This way, you can feel your best—so that you can be your best—at the table and beyond.

WARMING UP YOUR DIGESTIVE ENGINE

Eating and sharing delicious and nourishing food is key to living a Genius Life, but if you aren't properly digesting your food, you're being short-changed! Great digestion is nonnegotiable when it comes to feeling good inside and out. Not only does a well-tuned digestive furnace ensure that you're able to reap the myriad nutrients in the foods you've bought and prepared, but it minimizes or altogether eliminates digestive inconveniences like bloating, gas, or worse.

This may come as a surprise, but great digestion begins before you sit down. When stressed, digesting food is the last thing our bodies want. Historically, stress would have come from a source of danger—an enemy or a predator, for example. For this reason, stress, no matter where it comes from, profoundly reshuffles your body's priorities, and breaking down and assimilating nutrients—which accounts for about 10 to 20 percent of your body's energy use at rest—becomes an afterthought. In fact, as a survival tactic, blood actually leaves the digestive system of a stressed person and floods their muscles and vital organs instead.

CHECK YOUR POO

One reliable way to know whether you're properly digesting your food is to regularly look at your, ahem, output. Without going into too much detail (this is a cookbook after all), loose or oily stools and undigested food particles are usually a sign that something's awry somewhere along the canal. Don't be shy; in certain parts of the world, people take pride in their BMs. Dutch toilets, for example, have shelves in the bowls—an odd first-time experience for the uninitiated. This is thought by many to encourage poo-spection. Perhaps proper digestion (and happy number twos) is why the Netherlands consistently ranks among the happiest countries in the world. *Proost!*

Stress can also increase digestive motility. Ever get nervous before a job interview, presentation, or performance and get the runs? Or perhaps you have irritable bowel syndrome (IBS) and notice that your flare-ups are more common when you're stressed? For all of us, anxiety, stress, and even on-the-go eating can all lead to poor digestion. Once seated at the table (you *are* sitting, right?), take a few moments to breathe deeply and through your nose—five to ten slow, deep breaths is fine. Nose breathing helps activate your parasympathetic nervous response, aka your "rest and digest" state—the antidote to stress and the key to a happy tummy.

While you're breathing, allow your gaze to soften. Vision is our dominant mode of sensation, and visual input has a huge impact on the state of your nervous system. When stressed, an animal becomes hypervigilant, narrowing its vision to the threatening object. When grazing, however, animals keep a wide-open, panoramic view. According to my friend and Stanford professor Andrew Huberman, who studies the neurobiology of vision, by dialing back your field of vision to take in your surroundings, you actually can "force" your nervous system to calm down. Get into the habit of practicing this breathing and visual routine before every meal.

Nose breathing has another important benefit for digestion—it increases the production of a gas called *nitric oxide,* which does not occur with mouth breathing.* Nitric oxide relaxes the blood vessels around your body, reducing your blood pressure and increasing blood flow to your digestive tract. It also boosts your cells' sensitivity to the hormone insulin, which helps you to more effectively metabolize carbohydrates and sugar. Low nitric oxide has been correlated with higher risk of heart disease and type 2 diabetes.

There's another important way that we create nitric oxide: when chewing our food. When you chew foods that contain a compound called *nitrate,* bacteria in your mouth create nitric oxide, which you then absorb soon after swallowing. Plants tend to be rich in nitrates, with the most concentrated sources being beets, celery, and dark leafy greens like kale, spinach, and arugula (arugula contains the highest concentration, calorie for calorie). The key to optimizing nitric oxide release—and reaping all the benefits to your cardiovascular system—is to allow your bacteria time to do their job, by chewing slowly, about twenty to thirty seconds per mouthful.

---

* What makes your nasal airway unique is that the blood vessels that line your paranasal sinus (the cavity around your nose) are rich in an enzyme called *nitric oxide synthase,* which creates nitric oxide!

HOW MOUTHWASH CAN HURT YOUR HEALTH

A healthy mouth can contain upward of ten billion bacteria, which survive off the remnants of the foods we eat. Some of these bacteria can create problems like bad breath and dental caries (cavities), but others are necessary for good health, generating important compounds like nitric oxide. Unfortunately, two hundred million Americans use antiseptic mouthwash or fluoridated toothpaste every day, which kill off not just the bad bacteria but the good bacteria as well. Nathan Bryan, a preeminent expert and pioneer in our understanding of nitric oxide metabolism, told me, "Without these bacteria, we do not get all of the cardiovascular benefits of eating a healthy diet." Oral bacteria even recycle the nitric oxide we naturally create when we exercise, and studies have shown that mouthwash after exercise can negate its blood pressure–lowering effects. It's no wonder then that a surprising number of studies have found that regular users of mouthwash have increased risk for hypertension, heart disease, and type 2 diabetes—they are chronically nuking the bacteria that make this important, health-promoting pathway possible! Instead, practice good oral hygiene by avoiding refined grains and added sugars, and save the mouthwash for when it's truly medically necessary. Floss nightly, and stick to fluoride-free toothpastes with xylitol, a natural anticavity agent.

Chewing slowly has other benefits. Remember those beneficial plant defense compounds from page 5? Their precursors are often held in separate cellular compartments that unite upon injury to the plant's cells. Think microscopic Power Rangers for your health that combine to create the Megazord. In the case of cruciferous veggies like kale, broccoli, and Brussels sprouts, chewing yields sulforaphane, created by the union of glucoraphanin and myrosinase.* Sulforaphane is a putative cancer fighter and

---

* Myrosinase, one of the two compounds needed to create sulforaphane, is destroyed by heat, so cooked crucifers do not fully create sulforaphane. One workaround is to sprinkle mustard seed powder on your cooked veggies. Mustard greens are themselves cruciferous vegetables and their seeds are replete with myrosinase. A little dash on top of your cooked broccoli or Brussels sprouts (or any other cruciferous veggie) and you're back in cancer-fighting, brain-protecting business!

neuroprotectant, and it spurs your body to produce protective molecules like glutathione, a master detoxifier and antioxidant.[41] Alliums like garlic, leeks, and onions are known for producing allicin, another anticancer agent which may also help fight infections. The key to maximizing production of these powerful plant compounds is simply to slow down while chewing. Go, go, Power Rangers!

Chewing more slowly can also help you feel better. Here's how: starches are partially digested by enzymes in your saliva, ensuring their proper absorption. Malabsorption of these carbohydrates can cause bacteria in your gut to ferment them instead, leading to gas and bloating. Chewing also stimulates digestive enzymes like pepsin and lipase, which digest proteins and fat, respectively. With so many important digestive processes initiated upon chewing, it's no wonder that people who wolf down their food experience more bloating, indigestion, and other maladies of the midsection.

The takeaway here is, if you're the type to "inhale" your food, take a breath instead, and just slow down.

CHEW SLOWLY, LOOK BETTER NAKED

What could the speed at which you eat have to do with your waistline? Fast eating is associated with obesity, and one hypothesis is that hormones involved in hunger and satiety get short circuited with rushed eating. But the speed at which you chew might also affect the thermic effect of feeding, aka the amount of calories you burn via digestion. According to a 2015 study published in *The International Journal of Food Sciences and Nutrition*, women burned significantly more calories when they ate the same meal over a leisurely span of 15 minutes compared to when they ate it over a rushed 5 minutes. If that news wasn't sweet enough, get this: slowing down can also keep your blood sugar from spiking too high when eating a high-carb meal—no surprise since chewing prepares your body to release insulin (among other things), the hormone that lowers your blood sugar.

MAXIMIZING YOUR STOMACH ACID

You may not realize it, but acids are key to many good things in life. They help to balance out salt and fat in your culinary creations. They provide the "zing" in delicious citrus fruits like lemons, grapefruit, and oranges. Vinegar *is* an acid, which lends kick to everything from salads to roasted veggies and meats. And acids add complexity to movie plotlines: *Don't shoot the alien, it bleeds acid!* Of all, the most important acid may be the one your stomach produces.

Stomach acid gets a bad rap; we've all had heartburn at one point or another, straining our relationship with this mysterious, but vital, solution. Without your stomach acid, you'd die; it breaks your food down into its constituent nutrients so that you can absorb them and thrive. Sadly, over one hundred million people take prescription stomach acid–blocking medications called proton pump inhibitors (PPIs) every year. Many more take acid-neutralizing chewables or over-the-counter PPIs. While there are certainly medical conditions that warrant temporary use of these drugs, the repercussions of long-term acid-blocking are dire (I'll offer other ways to reduce reflux shortly).

Folate and vitamin B_{12} are two nutrients that require stomach acid to be absorbed. They help with cellular detoxification and brain function, among other things. Important minerals like calcium, potassium, magnesium, and zinc also need acid to be absorbed. Each of these minerals plays important roles in feeling good and being healthy. Magnesium, for instance, helps repair damaged DNA and maintains a healthy heart rhythm. Zinc keeps your immune system strong. And, of course, protein satiates your hunger and keeps *you* strong (among other things). Protein *requires* digestive acid to be broken down. Without it, otherwise beneficial proteins can cause problems from bloating to allergies.

We also require stomach acid to generate nitric oxide. Remember that powerful blood pressure–lowering gas we create in our airways when we breathe through our nose and chew nitrate-rich veggies like arugula and beets? Stomach acid plays a role here as well, and reduced stomach acid can decrease nitric oxide, thereby diminishing the cardiovascular benefits of eating a healthy, nitrate-rich diet. Perhaps these reasons (among others) are why long-term use of PPIs has been linked with cardiovascular disease along with other conditions for which high blood pressure is a risk factor.[42]

Having healthy levels of stomach acid is also important to keep your small intestine free of bacterial overgrowth. Bacteria don't love swimming in acid—and really, who does? When the normally acidic environment of the upper gastrointestinal tract is diminished, it allows bacteria to float and linger where they shouldn't. This can lead to a condition called *small intestinal bacterial overgrowth,* or SIBO. SIBO symptoms aren't pleasant; they include uncomfortable gas and diarrhea. The key to preventing bacterial overgrowth is to encourage healthy stomach acid levels, not reduce them.

Here are some tips to optimize your stomach acid to reap all the comfort and health benefits my dishes offer:

- **Slow down while chewing.** A full 30 to 50 percent of stomach acid is stimulated before you even swallow (the cephalic, or "head," phase of digestion).[43] Chew slower!

- **Sip, don't guzzle.** Sipping water, warm beverages, and soups can all improve digestion. But excess water while eating can raise the pH of your stomach, thereby reducing its acidity.[44] Though a small effect and temporary, this lends credence to what Ayurvedic texts have been saying for centuries: don't dilute your digestive juices with excess fluids while eating.

- **Consider vinegar.** Vinegar is an acid that some believe can help with digestion. Don't drink it undiluted; it can ruin your tooth enamel and irritate your throat. On page 56 I offer a refreshing vinegar beverage recipe that can help ease digestion.

- **Lose weight.** If you're overweight and experiencing reflux, losing weight can seriously help. That's because the excess weight puts pressure on the junction between your stomach and esophagus, and dropping weight eases the pressure.

- **Cut out refined carbs.** Studies suggest that lower carbohydrate diets can improve symptoms of reflux. You'll get all the carbs you need to feel great in this cookbook, sans the refined ones.

- **Don't eat for two to three hours before bed.** Circadian biology (your body's relationship with time) dictates that food is best consumed two to three hours before bed. This ensures that by the time you lie down, the food you've consumed has already left your stomach. (You can learn much more about your body's circadian rhythms in my book *The Genius Life*.)

- **Consider stomach enzymes or betaine HCl.** As we age, we tend to produce less acid and digestive enzymes. Betaine HCl is supplemental hydrochloric acid (betaine is naturally found in foods like beets and spinach), and digestive enzymes like pepsin (for protein) and lipase (for fat) can help enhance digestion if you've recently taken acid blockers or are experiencing digestive issues.[45]

EAT TO BOOST SEXUAL PERFORMANCE

Can your eating habits set the stage for knocking boots? You bet. As you've discovered, nasal breathing and chewing your food slowly all lead to increased nitric oxide release, which supports blood vessel relaxation all over your body, including you-know-where. In fact, low nitric oxide release is related to erectile dysfunction, which can be an indicator of subclinical heart disease. Acid-blocking medications (and antiseptic mouthwash, which I've covered on page 87) can each pump the brakes on your nitric oxide pathway. So, ditch the drugs, breathe through your nose, and chew your food slowly. The time to peak nitric oxide release is about ninety minutes . . . the perfect interval to enjoy the meal, share dessert, and put on your sexy time playlist of choice. Thank me later!

I hope this section has made clear the methods essential for reaping all there is to gain from healthy and delicious cooking. Now, without further ado, the recipes!

12 GENIUS COOKING TIPS

1. Read your recipes in their entirety first!

Ever make it halfway through a recipe only to realize you've neglected to pick up a necessary ingredient? I've done it! The solution: don't be hasty. Read through the entire recipe first so that there are no surprises.

2. Have patience.

In today's on-demand world, we want our food quickly, and still expect it always to be delicious. The art of low and slow cooking has become a relic. Some dishes require time; for example, my Insanely Crispy Gluten-Free Buffalo Chicken Wings (page 194), Perfect Ribs (page 202), or Bone Broth Beef Stew with Purple Sweet Potatoes (page 191). Some meats (tough cuts in particular) truly shine only when cooked low and slow, when the connective tissue has been given time to melt away. Many veggies need time to caramelize. Don't rush the process.

3. Thaw plastic-wrapped foods in the fridge or cold water.

Utilizing frozen foods is a great way to minimize food waste and cut down on grocery costs. Ideally, you'd like to thaw foods in the refrigerator overnight, but for more immediate use, place the item in cold water. The water will still thaw the food, but by avoiding putting plastic-wrapped foods in hot water, you minimize the chances of transferring potentially harmful plastic chemicals into your food.

4. Don't overuse fats and oils.

I'm a proponent of using the right fats and oils for optimal health, but there is still a limit to how much can be used before the calories start adding up! Use as much as you need to cook, but no more. Some foods don't even need added fat in the pan, like when cooking a burger patty, since ground beef already has plenty of fat of its own. When using fats like extra-virgin olive oil or butter as a dressing or spread, the amount to use depends on your goals; I like to stick to a one- to two-tablespoon cap on added fat as a general rule of thumb.

5. Clean as you go.

Cooking is fun, but the prospect of cleanup can be agonizing unless you do it as you go. Thankfully, not much effort is involved. In between recipe steps, throw out scraps, wipe down your countertops with a towel (paper or cloth), and soak pots and pans before the food dries and becomes caked on. It'll save you a headache later.

6. Preheat!

Make sure your oven is fully preheated. The reason for this is simple—if you allow your oven to heat up with your food in it, it could lead to uneven cooking. Same goes for pans. Plus, when your pan is hot, your food is less likely to stick. And a hot pan will give you that delicious sear you're looking for, whereas a tepid pan will overcook your food before you get to the sear you want.

7. Master the art of salting.

I loathe bland food. So do most chefs. In fact, ask any chef what the biggest mistake most home cooks make is and they'll likely answer: undersalting. While it's true that it's safer to undersalt than to oversalt (you can't remove salt once added, but you can always add it later), mastering the salting process is crucial to making flavors pop. Don't feel guilty about salting your food to make it delicious—seriously! (Revisit page 51 to understand why.) Even sweet foods need a little bit of salt to shine (see my Chocolate Blueberry Clusters on page 270 to see salt and sweet in action). Keep in mind that sauces like coconut aminos, tamari and fish sauce, prepared broths, and canned foods generally will add plenty of salt on their own, so adjust your seasoning accordingly.

8. Taste often!

There are no points to be won by not tasting your food until it's done. Eliminate the guesswork by tasting your food and tasting frequently, through every step (except, of course, if unsafe to do so). This is how all chefs cook, and it allows you to learn how flavors (and textures!) develop over time.

9. **Learn how to properly use a knife.**

As someone who's largely self-taught, I'm still mastering this one. Learning how to properly use a knife is key to being efficient in the kitchen while also preventing bloodshed. When cutting or chopping, always hold the item with your fingers pointed down, nails parallel to the blade, so that you can't inadvertently slice them. For a visual guide, you can always head to YouTube and search for "basic knife skills."

10. **Enjoy yourself.**

Cooking should first and foremost be fun. It's sort of like karaoke: whether you're a terrible singer or a budding pop star, if you're not having fun with the mic, no one's having fun watching you. Same with cooking; whether your meal turns out divine or disgusting, as long as you've had fun and have kept a positive attitude about it, your dinner dates will laugh along with you.

11. **Cook with friends.**

One of my favorite things to do is to cook with friends. Don't get me wrong, I love cooking for my brothers, but . . . they're my brothers. Inviting friends over deepens social bonds while offering your comrades a powerful gift: food. Humans have been bonding over food for millennia, and we're not stopping anytime soon. Make cooking a collaborative process by delegating parts of the recipes you're following.

12. **Play music while you cook and serve.**

Jean-Michel Basquiat said, "Art is how we decorate space, music is how we decorate time." Have music that you like to play while you're cooking, serving, and eating. It elevates the experience, adds a rhythm to your activities, and lifts the mood. While eating, aim for music that calms you to encourage the "rest and digest" state—save the hard rock for your workouts. Me? I love to play anything from French jazz to mellow electronic music.

PART TWO

RECIPES FOR A GENIUS LIFE

CHAPTER 3

BREAKING THE FAST

Whether you're an early or late riser, the following recipes are perfect for your first meal. My intent in creating them was to delight your palate and wake you up while providing ample satiety and protein to keep you feeling strong all day long.

The World's Best Blueberry Smoothie That Actually Tastes Like Blueberries

I am almost always let down by blueberry smoothies, as they rarely taste like their namesake. That's why I've done things a little differently by leaving out the over-powering ingredients (e.g., banana) and playing up the tanginess of the blueberries with citrus and a little nondairy yogurt. I think you'll find the results berry, berry delicious. Consume as is, or add a scoop of vanilla-flavored whey protein isolate.

Serves 1

1 cup frozen blueberries
½ avocado, flesh scooped out
½ small tangerine, peeled and segmented
1 cup plain unsweetened coconut yogurt*

Zest of ½ lemon
1 tablespoon ground flaxseeds
Coconut milk or water, as needed

* If you enjoy dairy, you can substitute plain dairy Greek yogurt for added protein.

In a blender, pulse the blueberries, avocado, tangerine, yogurt, and lemon zest until combined (it'll be thick). Add the flaxseeds and, with blender running, drizzle in coconut milk, a little at a time, until you've reached the consistency you prefer. Serve immediately.

Beef Breakfast Sausage Patties

If you're someone who likes sausage for breakfast (or sausage gravy, for that matter), this is a healthy twist on the preservative-laden classic many of us have become used to. And yes—you can use this sausage mix as a base for gravy for a Sunday-morning treat.

Serves 4

1 pound ground beef
1 tablespoon rubbed sage
1 teaspoon salt
1 teaspoon onion powder

1 teaspoon garlic powder
1 teaspoon ground black pepper
1 teaspoon paprika
1 teaspoon monk fruit sweetener

With clean hands, crumble the beef into a large bowl. Add the remaining ingredients and work them in until fully incorporated.

Fill a small bowl with water. Shape the meat mixture into 8 ½-inch patties, wetting your hands in between patties (it's easier to shape them with wet hands).

Heat a skillet over medium-high heat and, working in batches, add the patties and cook for 2 to 3 minutes per side, until nicely browned on both sides and 140–150 degrees in the middle.

If you're not cooking the patties right away, you can freeze them. Store in the freezer for up to 3 months, and cook from frozen for 3 to 4 minutes per side.

Grain-Free Blueberry Orange Pancakes with Coconut Cream

There's a reason blueberry pancakes are a brunch staple—it's because they're so satisfying. I find the addition of orange zest and a tiny bit of cinnamon makes the pancakes taste almost like blueberry pie.

Serves 4

5 large eggs

1 cup full-fat canned coconut milk, plus more as needed

Zest and juice of ½ orange

1 tablespoon pure vanilla extract

2 tablespoons monk fruit sweetener

2/3 cup cassava flour (Otto's is a popular brand)*

1/3 cup coconut flour

1/4 cup almond flour

3/4 teaspoon baking powder

1 pinch salt

1/4 teaspoon ground cinnamon

Coconut oil, ghee, or avocado oil, for cooking

1 cup fresh blueberries

1 (13.66-ounce) can unsweetened coconut cream, refrigerated for 24 hours

1 tablespoon powdered Swerve (optional)

1 tablespoon hemp hearts (optional)

*If you can't find cassava flour, add an extra ⅓ cup almond flour and 2 extra tablespoons coconut flour for a similar flavor.

Chill a metal mixing bowl or the bowl of your stand mixer in the fridge.

In a blender or a large bowl, blend or whisk the eggs, coconut milk, orange zest and juice, the vanilla, and monk fruit until smooth.

In a small bowl, whisk together the cassava, coconut, and almond flours, the baking powder, salt, and cinnamon.

Add the wet ingredients to the dry ingredients in the blender or bowl and blend or whisk until smooth. If the mixture looks dry, add coconut milk 1 tablespoon at a time until you have a traditional pancake batter consistency.

Preheat the oven to 200°F.

Heat a large skillet over medium heat. Add 1 tablespoon oil. Pour ¼-cup portions of batter onto the skillet, leaving space for the pancakes to expand, and sprinkle each pancake with blueberries.

Cook for 3 to 4 minutes, then flip and cook for an additional 3 to 4 minutes, until pancakes are golden brown on each side, and no raw batter shows when pierced in the center with a sharp knife. As the pancakes are cooked, transfer them to a sheet pan in the oven to keep warm.

When all the pancakes are cooked, open the chilled can of coconut cream and scoop the thick cream off the top of the can into the chilled bowl. Add the Swerve and whip with an electric mixer for 3 to 5 minutes, until fluffy. Place the pancakes on plates and serve with the whipped coconut cream and sprinkle hemp hearts over the top.

"Cheesy" Baked Eggs with Broccoli

There's something so classic about the combination of broccoli and cheese—and if you're big on a cheesy broccoli quiche for breakfast, this dairy-free recipe is perfect for you. It's a great brunch on its own or served with a side of salmon skin "bacon" (see page 111).

Serves 4

3 tablespoons avocado oil, divided

2 cups ¼–½ inch broccoli florets (½ large head)

1 teaspoon salt

3 tablespoons nutritional yeast, divided

2 tablespoons unsweetened coconut cream, divided

Zest and juice of 1 lemon

4 eggs

Hot sauce (optional)

Preheat the oven to 375°F. Use 1 tablespoon of avocado oil to lightly oil four 4-ounce oven-safe ramekins.

Heat a large skillet over medium-high heat. Add remaining 2 tablespoons of the oil, the broccoli, and salt and cook until the broccoli is tender at the very edges. Add 2 tablespoons of the nutritional yeast, 1 tablespoon coconut cream, and the lemon zest and juice. Cook, stirring, until well combined and the broccoli is vibrant green, tender at the edges but still firm.

Remove from the heat, divide the broccoli among the prepared ramekins, and crack an egg over each.

In a small bowl, whisk together the remaining 1 tablespoon nutritional yeast and 1 tablespoon coconut cream and pour it over the eggs. Bake for 15 to 20 minutes, until eggs are set to your liking. Serve with hot sauce.

Herb and Avocado Scrambled Eggs

Scrambled eggs with herbs is a classic for a reason—and this simple scramble takes it over the top by keeping the eggs light and creamy and cheesy-tasting without being loaded with dairy. It's perfect for a quick weekend breakfast (or as part of a hearty brunch), and even better with my Olive Oil–Poached Salmon (page 228).

Serves 2

4 eggs
2 tablespoons minced fresh dill
2 tablespoons minced fresh parsley
2 tablespoons minced fresh cilantro
3 tablespoons nutritional yeast
1 tablespoon unsweetened coconut cream
2 tablespoons avocado oil

1 shallot, minced
1 jalapeño, minced (omit if your tastes run mild)
½ teaspoon salt
1 avocado, diced
¼ cup minced fresh basil

In a large bowl, whisk together the eggs, dill, parsley, cilantro, nutritional yeast, and coconut cream.

In a small skillet, heat the oil over low heat. Add the shallot, jalapeño, and salt and cook until the vegetables are fragrant and tender, about 5 minutes. Add the eggs and continue to cook, stirring frequently, until the eggs are creamy and just beginning to set, about 7 minutes.

Fold in the avocado, cook until the eggs are set to your liking, then fold in the basil and serve.

Olive Oil–Poached Salmon "Benedict"

When I've got time to make a big brunch in the morning, there's nothing better than a benedict. While I'm down with the occasional Canadian bacon, I wanted to see what would happen if I replaced it with something even better for you—olive oil–poached salmon. And I've nixed the English muffins in favor of nutrient-dense (and damn delicious) hash brown–style sweet potato cakes.

Serves 2

1 cup shredded sweet potato
1 tablespoon tapioca starch
½ teaspoon salt
1 teaspoon minced fresh rosemary
2 tablespoons extra-virgin olive oil, divided

4 (2-ounce) slices Olive Oil–Poached Salmon (page 228)
2 eggs
1 recipe Easy Olive Oil Hollandaise (recipe follows)
Minced fresh dill

In a medium bowl, toss together the sweet potatoes, tapioca starch, salt, and rosemary.

Preheat the oven to 350°F.

Heat 1 tablespoon of the oil in an oven-safe skillet over medium-high heat.

Shape the sweet potato mixture into 2 English muffin–sized cakes and place in the skillet.

Cook for 2 to 3 minutes, until golden brown on the bottom, then remove from the heat.

Put the salmon on top of the sweet potato in an even layer, then carefully crack an egg on top of each and drizzle with the remaining 1 tablespoon oil.

Transfer to the oven and bake until the whites are set, 8 to 10 minutes (this is the perfect time to make your hollandaise). Transfer each "benedict" to a serving platter and drizzle with hollandaise. Top with dill and serve.

Easy Olive Oil Hollandaise

Serves 2

⅓ cup mild-tasting extra-virgin olive oil
1 egg yolk
2 tablespoons warm water, divided

1 teaspoon paprika
Juice of ½ lemon
1 pinch salt

In a small saucepan, gently heat the oil to 120°F. Remove from the heat.

In a blender, combine the egg yolk, 1 tablespoon of the warm water, the paprika, lemon juice, and salt and blend until smooth.

With the blender running, slowly stream the oil into the mixture through the hole in the lid—it will thicken up pretty quickly. If it's too thick, thin with additional water.

Serve over Olive Oil–Poached Salmon "Benedict" (page 108) or with grilled asparagus for an easy side dish.

Perfect Poached Eggs

Knowing how to poach an egg is a skill that—like playing guitar or practicing yoga—you're likely to appreciate having at least a few times in life. It involves gently cooking a cracked egg in hot water. The beautiful thing about egg poaching is that it requires no added oil and gives your eggs a delicious taste and texture. When done right, the yolk stays nice and custard-like, maintaining all its healthy fats. And the whites take on a slightly acidic flavor from the vinegar, which complements any dish it accompanies. I often top these with a dash of flake salt and eat with smoked salmon. This recipe is for one serving, but you can batch cook as many as you like using the same water.

Serves 1

2 eggs   2 tablespoons distilled white vinegar

Fill a medium-sized, straight-sided saucepan with 4 to 5 inches of water, enough to cover the eggs completely. Add the vinegar and bring to a boil, then dial the heat down to maintain a very low simmer.

Crack one egg into a shallow bowl, being careful not to break the yolk.

Once the water has calmed down, gently slide the egg into the water. It helps if you put the edge of the bowl into the water. Repeat for the second egg. Set a timer for 4 minutes and 15 seconds.

When the timer goes off, using a slotted spoon, remove the eggs from the water, again being careful not to break the yolks. Gently pat the eggs with a paper towel before serving.

Crispy Salmon Skin "Bacon" and Eggs

Whenever I make a recipe that calls for skin-off salmon, I buy skin-on and save the fat and collagen–rich skin to make a healthier "bacon." Just store it in your freezer and let it thaw completely before using. You can use salmon skin "bacon" anywhere bacon is called for (on top of a salad is one great place). If you like a smokier bacon, try adding a drop or two—no more—of liquid smoke to the marinade.

Serves 2

¼ cup coconut aminos
1 teaspoon fresh lemon juice
1 teaspoon monk fruit sweetener
Skin from 4 salmon fillets (save the flesh for another recipe)

2 tablespoons avocado oil or melted ghee, divided
1 shallot or ¼ red onion, minced
4 eggs
¼ cup minced fresh dill
½ teaspoon salt, or to taste
2 tablespoons sesame seeds

In a small saucepan, combine the coconut aminos, lemon juice, and monk fruit. Cook over medium-low heat for 3 to 5 minutes, until the monk fruit dissolves. Let cool to room temperature, then pour into in a sealable dish. Add the salmon skins, submerging them completely. Let rest in the refrigerator for at least 3 hours (or as long as overnight). Remove from the refrigerator 30 minutes to 1 hour before cooking.

Preheat the oven to 275°F. Line a sheet pan with parchment paper and brush with 1 tablespoon of the oil.

Carefully dry the salmon of any excess marinade and lay the salmon skins flat on the sheet pan in a single layer.

Bake for 40 minutes to 1 hour, checking every 10 minutes to make sure the sweetener doesn't burn, until the skin looks crisp and dry. Remove from the oven.

While salmon is baking, heat the remaining 1 tablespoon oil in a small skillet over medium-high heat. Add the shallot and cook, stirring occasionally, for 2 to 3 minutes, until tender.

Meanwhile, in a large bowl, whisk the eggs with the dill and salt.

Add the egg mixture to the pan and cook, stirring frequently, until the eggs have reached your desired level of scramble. Sprinkle with the sesame seeds and serve with the salmon skin "bacon."

Baked Eggs in Sweet Potato "Boats" with Herb Relish

One of my favorite places to eat in LA is the hot bar at a natural foods store called Erewhon. For breakfast, they serve a fried egg and sweet potato dish that inspired this one. This recipe is a great way to use up extra sweet potatoes—I often like to bake a few at the start of the week and keep them in the fridge just for this recipe. The herbs and capers brighten up the richness of the eggs, and the raisins add a surprising zing.

Serves 4

2 medium-large fully baked sweet potatoes
2 tablespoons extra-virgin olive oil
4 eggs
½ teaspoon salt
¼ cup minced fresh parsley
¼ cup minced fresh cilantro

2 tablespoons capers
2 tablespoons raisins, roughly chopped
1 small shallot, minced
Juice of 1 lemon
About 1 tablespoon red wine vinegar

Preheat the oven to 375°F.

Slice each baked sweet potato in half lengthwise and scoop out enough of each inside to make room for an egg. Brush with the oil inside and out.

Arrange the sweet potatoes in an baking dish and crack an egg into the hollow created in each sweet potato. Sprinkle with the salt, then transfer to the oven. Bake for 8 to 10 minutes, until the eggs are set to your liking.

Meanwhile, make a relish by combining the parsley, cilantro, capers, raisins, shallot, and lemon juice in a small bowl. Taste and add vinegar a little at a time until just tangy enough for your taste buds.

Serve each egg "boat" topped with relish.

CHAPTER 4

SMALL BITES

These dishes are delicious sides or, paired with a protein, main courses. They're also wonderful for parties or to bring to potlucks. The focus in these dishes (aside from teasing your taste buds) is to provide your body with important phytochemicals that strengthen you from the inside out.

Grain-Free Greens Fritters with Avocado Dipping Sauce

We've all been there: you bought more produce than you could comfortably eat within a given period, and you're worried it's going to go bad. Whenever I end up with a surplus of greens, I turn them into fritters—it's also a great way to use up any slightly wilted greens that you don't want to waste but don't want to eat raw. These crispy, salty, flavorful fritters are balanced perfectly by a tangy, creamy, avocado-based sauce.

Serves 4

1 cup baby spinach leaves, roughly chopped and packed

1 cup fresh kale leaves, roughly chopped and packed

1 cup chard leaves, roughly chopped and packed

1 teaspoon salt, divided, plus more as needed

1 small shallot, very finely chopped

1 clove garlic, minced

1 teaspoon ground black pepper

½ teaspoon ground cumin

½ teaspoon paprika

½ teaspoon dried oregano

1 egg

4 tablespoons cassava flour, plus more as needed

2 ripe avocados, flesh scooped out

Juice of 1 lemon

¼ cup fresh parsley, chopped

¼ cup fresh dill, chopped

1 tablespoon sesame oil

Dash of coconut aminos or tamari

3 tablespoons avocado oil

In a food processor (or by hand with a very sharp knife), pulse the spinach, kale, and chard until very finely chopped. Remove from the food processor and sprinkle lightly with ½ teaspoon salt. Let rest for 10 minutes, then place between paper towels and squeeze to wring out excess moisture (you can also do this with a cloth towel, but be aware that it may stain).

Transfer the wrung-out greens to a large bowl and add the shallot, garlic, black pepper, cumin, paprika, oregano, egg, and 2 tablespoons of the cassava flour. Mix to combine, then test if the mixture will form nice fritters that hold together when squeezed. If it's too dry, add a small splash of water. If it's too wet, add cassava flour, a couple teaspoons at a time, until the mixture binds nicely.

Using your hands or an ice cream scoop, shape the mixture into 8 to 10 small balls. Flatten slightly, so they're somewhere between a ball and a patty shape. Place on a plate or sheet pan and chill in the freezer for 5 to 10 minutes or in the refrigerator for at least 30 minutes or up to overnight.

While the fritters are chilling, make the sauce: in a food processor or blender, combine the avocados, lemon juice, parsley, dill, sesame oil, and coconut aminos and process until smooth. Transfer to a serving bowl and cover the surface with plastic wrap to prevent browning. Chill until ready to serve.

Heat the avocado oil in a large skillet over medium heat. Carefully place the fritters in the skillet, spacing them evenly, and cook for 3 to 5 minutes, until the bottom side is golden brown. Flip, reduce the heat to low, and cook for an additional 5 to 7 minutes, keeping a close eye, until the bottoms brown nicely but don't burn and the fritters are cooked all the way through. Remove from the heat, sprinkle with the remaining salt, and serve with the avocado sauce.

Romaine and Broccoli with Sesame Caesar Dressing

I love a classic Caesar dressing—and this sesame-heavy twist omits the cheese in favor of creamy, nutty tahini for a dairy-free but equally creamy dressing. Raw broccoli adds a bit of sweetness, and the hemp hearts—aside from lending some additional protein—are perfect for a bit of crunch. Enjoy as is, or add a piece of grilled fish, chicken, or shrimp for a hearty meal.

Serves 2 to 4

1 head broccoli, with stalk

1 heart romaine lettuce, limp outer leaves removed

3 tablespoons hulled hemp hearts

3 tablespoons toasted sesame seeds

1/2 teaspoon salt

1 teaspoon ground black pepper

2 tablespoons extra-virgin olive oil

1 tablespoon Dijon mustard

1/4 cup tahini (sesame seed paste)

Juice of 1/2 lemon

1 small clove garlic, peeled

1 dash coconut aminos or tamari

1 oil-packed anchovy fillet (optional but delicious—and not at all fishy!)

To prepare the broccoli, trim the woody ends off the stalk and peel the stem. Chop the stalk into small matchsticks and set aside. Chop the remaining broccoli into very small florets. Tear or slice the lettuce into bite-sized pieces.

In a large bowl, toss the prepared broccoli with the lettuce and set aside.

In a small bowl, combine the hemp hearts, sesame seeds, salt, and black pepper. Sprinkle over the salad.

In a blender, combine the oil, mustard, tahini, lemon juice, garlic, coconut aminos, and anchovy, if using (try it once—SO worth it!). Toss with the greens and serve.

Watercress, Avocado, and Grilled Fruit Salad with Toasted Almond Dressing

Tangy citrus and creamy avocado are perfect partners in salad—and the addition of tropical fruit balances out the peppery bite of watercress. I've used walnut oil, made by simply pressing walnuts, as a finishing oil in this salad. If you don't have walnut oil, that's okay—use sesame or olive oil instead for a slightly different but equally delicious flavor.

Serves 4

Extra-virgin olive or avocado oil

1 medium-ripe mango, pitted, peeled, and sliced 1-inch thick

1 peach, pitted and sliced 1-inch thick

¼ pineapple, peeled and sliced 1-inch thick

1 grapefruit, peel on, cut into 1-inch-thick slices

¼ cup minced fresh mint

4 cups watercress

2 avocados, sliced

1 small red onion, minced

½ cup toasted almonds

2 cloves garlic, peeled

2 tablespoons walnut oil

Juice of ½ lemon

1 teaspoon red wine vinegar

1 teaspoon fresh thyme leaves

Heat a grill (or a large skillet) to medium-high and brush the surface with olive or avocado oil.

Grill (or sear in the pan) the mango, peach, pineapple, and grapefruit until lightly charred on the outside, 2 to 3 minutes per side. Remove from the heat and let cool to room temperature. Chop into bite-sized pieces, place in a large bowl, and toss with the mint.

Arrange the watercress on a large platter or in a big bowl. Top with the avocados, onion, and grilled fruit.

To make the dressing: In a food processor, combine the almonds, garlic, walnut oil, lemon juice, vinegar, and thyme and process until smooth and creamy. Thin with water to create a pourable texture if needed. Drizzle the dressing over the salad and serve.

Baked Almond "Breaded" Artichoke Hearts with Herb and Kale Pesto

Who doesn't enjoy breaded, fried foods, right? Well, this delicious "breaded" artichoke dish isn't fried . . . and it's not covered in bread crumbs. And yet somehow it still tastes like junk food, in the best possible way. It's spicy, salty, and crispy—perfect for dipping into a garlicky pesto. Be sure to keep extra napkins around!

Serves 4

4 cups frozen quartered artichoke hearts, thawed
¼ cup tapioca starch
2 eggs
¼ cup full-fat canned coconut milk
1 generous dash of your favorite hot sauce
2 cups almond flour
1½ cups nutritional yeast, divided
1 teaspoon paprika

1 teaspoon garlic powder
2 tablespoons avocado oil
1½ cups shredded kale leaves
½ cup fresh basil, roughly chopped
½ cup fresh parsley, roughly chopped
½ cup whole toasted almonds
2 cloves garlic, peeled
1½ teaspoons salt, divided
¼ cup extra-virgin olive oil

Preheat the oven to 425°F. Line a sheet pan with parchment paper, brush the paper with oil, and set aside.

Place the artichoke hearts and tapioca starch in a large bowl with a cover or a zip-top bag and shake until the artichoke hearts are well-coated.

In a separate bowl, whisk together the eggs, coconut milk, and hot sauce.

In an additional bowl or shallow dish, stir together the almond flour, ½ cup of the nutritional yeast, the paprika, and the garlic powder.

Carefully dip the artichoke hearts into the egg mixture, then into the almond mixture, and arrange on the prepared sheet pan in a single layer. Drizzle with the avocado oil, then transfer to the oven and bake for 10 to 15 minutes, turning occasionally so as not to burn the artichokes, until golden brown.

While artichokes are baking, make the pesto: In a food processor fitted with an "S" blade, pulse together the kale, basil, parsley, toasted almonds, and garlic cloves until a crumbly mixture forms. Add the remaining 1 cup nutritional yeast and ½ teaspoon of the salt. With the machine running, stream the olive oil into the food processor until well combined. If you like a thinner pesto, add a little more oil. Sprinkle the artichoke hearts with the remaining salt immediately before serving and serve with the dipping sauce.

Miso Coconut Simmered Veggies

This hearty braised veggie dish features affordable, easy-to-find vegetables, savory miso, and plenty of ginger to add sweetness and bite. Braising in coconut milk rather than broth balances out the bitter bite most greens have, making this dish accessible to most peoples' palates.

Serves 4

2 tablespoons sesame oil

1 medium red onion, thinly sliced

1 3-inch-diameter turnip, peeled and cut into bite-sized pieces

2 carrots, peeled and cut into bite-sized pieces

1 parsnip, peeled and cut into bite-sized pieces

1 zucchini, cut into rounds

2 tablespoons minced fresh ginger

2 tablespoons red or white miso paste (red preferred)

1 (15-ounce) can full-fat unsweetened coconut milk

1 teaspoon coconut aminos, plus more as needed

1 tablespoon rice vinegar

½ small head of napa cabbage, roughly chopped

1 bunch collard greens, stems removed, leaves thinly sliced

Heat the oil in a large, heavy-bottomed soup pot over medium-high heat.

Add the onion, turnip, carrots, parsnip, and zucchini and cook until the onion is tender and translucent, 5 to 7 minutes. Add the ginger and miso paste and cook until the ginger is fragrant and the miso paste is broken up throughout the dish, 2 to 3 minutes. Pour the coconut milk, coconut aminos, and vinegar into the pot, bring to a simmer, then reduce the heat to low, cover, and simmer for 15 to 20 minutes, until all the veggies are tender.

Add the cabbage and collard greens and cook uncovered for an additional 7 to 10 minutes, until the greens are wilted. Taste, add more coconut aminos as needed, and serve.

Lemon, Lentil, and Broccoli Soup with Apricots and Pomegranate

When I was growing up, my mom used to make a lentil soup like this one, and I just loved the dried apricots—they added sweetness and freshness to a meaty-tasting soup and really woke up my taste buds. I've updated this soup by adding broccoli along with pomegranate seeds and cilantro for an added punch of fresh-tasting flavor.

Serves 4

3 tablespoons extra-virgin olive oil
Zest and juice of 2 lemons
1 medium-sized yellow onion, minced
3 cloves garlic, minced
1 teaspoon ground cumin
½ teaspoon red chile flakes
1 cup dried red lentils

6 cups chicken bone broth or vegetable broth, divided
4 cups bite-sized broccoli florets (1 head)
1 cup minced dried unsweetened apricots
1 cup pomegranate arils
¼ cup minced fresh cilantro

Heat the oil in a large soup pot over medium-high heat. Add the lemon zest (save the juice for later), onion, garlic, cumin, and red chile flakes and cook until the onion is translucent and softened, 3 to 5 minutes.

Add the lentils and 3 cups of the broth and bring to a simmer. Lower the heat to medium-low, cover, and cook for 30 to 45 minutes, until the lentils start to fall apart and become creamy. You can help the process along by gently crushing them with the back of a fork.

Whisk in the remaining broth along with the broccoli and dried apricots. Raise the heat briefly, bring to a simmer, then reduce the heat to medium-low and cook uncovered for 10 to 15 minutes, until the broccoli is tender. Remove from the heat, add the lemon juice, pomegranate arils, and cilantro, and serve.

Vegan Carrot Noodle Mac and "Cheese"

I grew up loving boxed mac and cheese, which is delicious to a young palate but anything but a health food. Thankfully, creating a version that is equally indulgent isn't that difficult. If you've ever craved mac and cheese, this nut-based recipe will hit the spot . . . and for children, too! It's so delicious that most people won't miss the dairy. The "cheese" sauce is also great tossed over roasted broccoli (sorry, I just drooled all over my keyboard).

Serves 4

½ cup raw macadamia nuts, soaked in water for at least 8 hours or overnight

½ cup toasted pine nuts, soaked in water for at least 8 hours or overnight

¾ cup full-fat canned coconut milk

Juice of ½ lemon, plus more as needed

½ cup nutritional yeast, plus more as needed

1½ teaspoons salt, divided, plus more as needed

½ teaspoon smoked paprika

1 teaspoon Dijon mustard

1 clove garlic, peeled

1 pinch ground black pepper

4 cups carrot "noodles" from 4 large carrots

2 tablespoons avocado oil

1 medium red onion, thinly sliced

1 teaspoon red chile flakes

Drain the nuts of their soaking water.

In a food processor fitted with an "S" blade, combine the macadamia nuts and pine nuts, the coconut milk, and lemon juice and process until mostly smooth, scraping down the sides occasionally (this can take a few minutes). Add the nutritional yeast, ¾ teaspoon of the salt, the paprika, mustard, garlic, and black pepper and process until smooth. Taste and adjust the salt, nutritional yeast, and lemon juice to your desired level of "cheesiness."

Preheat the oven to 400°F.

In a large bowl, toss the carrot noodles with a ¼ teaspoon of salt. Let rest for 5 minutes, then pat dry with paper towels.

Heat the oil in a large skillet over medium heat. Add the carrot noodles, onion, remaining salt, and chile flakes and cook for 3 to 5 minutes, until the onion is tender.

Toss in the "cheese" sauce, then transfer to an oven-safe baking dish. Bake for 10 to 15 minutes, until bubbly, then serve.

Shrimp, Blueberry, and Arugula Salad

You'll notice quite a few chopped salads in this book—because, in my opinion, they're one of the best lunches for busy days, filling you up and checking off numerous nutritional boxes. When made with a heartier green, like arugula, you can make this salad in the morning (or even the night before) and have it stay perfectly fresh and crisp for lunch the next day.

Serves 4

1 lemon, thinly sliced (peel on)
3 cloves garlic, thinly sliced
1 tablespoon whole black peppercorns
1 teaspoon red chile flakes
1½ teaspoons salt, divided
1½ cups water
1 pound peeled, deveined shrimp (tail-on is OK)
4 cups arugula

1 small red onion, thinly sliced
2 cups fresh blueberries
1 small English cucumber, peeled and thinly sliced
Juice of 1 lemon
¼ cup extra-virgin olive oil
2 tablespoons rice vinegar
1 teaspoon Dijon mustard
2 avocados, thinly sliced

In a large saucepan, combine the lemon, garlic, peppercorns, chile flakes, ¾ teaspoon of the salt, and the water. Place over medium-low heat and bring to a simmer. Add the shrimp and cook for 5 to 6 minutes, until the shrimp is just cooked through. Remove the shrimp from the water (discard the seasonings) and immediately rinse in ice-cold water to prevent overcooking. Pat the shrimp dry with a paper towel.

In a large bowl, toss together the arugula, onion, blueberries, and cucumber. In a small bowl, whisk together lemon juice, oil, vinegar, and mustard until emulsified. Toss with the arugula mixture, top the salad with the shrimp and avocados, and serve.

Zucchini Noodles with Olive Oil, Garlic, and Chile Flakes

Sometimes simple dishes are best—and this ultra-simple, six-ingredient zucchini one is no exception. All that zucchini needed was plenty of good-quality olive oil, garlic, and pine nuts to go from simple to the star of your table. Try tossing in some Olive Oil–Poached Salmon (page 228) to take it over the top.

Serves 4

5 cups zucchini noodles (about 3–4 large zucchini, spiralized or cut into thin, spaghetti-sized strips)

1½ teaspoons salt, divided

¼ cup extra-virgin olive oil

6 to 8 cloves garlic, thinly sliced

1 teaspoon red chile flakes

¼ cup toasted pine nuts

In a large bowl, toss the noodles with ¾ teaspoon of the salt.

In a large saute pan, heat the oil and garlic over medium heat for 2 to 3 minutes, until the garlic is very light golden brown. With a slotted spoon, remove the garlic from the oil and reserve it for later. Make sure to remove all the garlic to prevent a bitter burnt garlic taste.

Carefully squeeze the zucchini noodles between 2 layers of kitchen towels to draw out moisture.

Turn the heat under the pan up to high. Add the zucchini noodles and chile flakes and cook for 3 to 5 minutes, until the noodles are tender and golden brown at the edges. Remove from the heat, top with the cooked garlic and the pine nuts, and serve.

Broccoli-Based Greek-Style Salad with Oregano Dressing

I'm such a big fan of Greek food; it's one of the cuisines I would regularly take my mom to enjoy. And every meal would start with sharing a Greek salad with her. In my opinion, the only thing better than the classic version is one with even more vegetables—in this case, with the addition of broccoli. Raw broccoli (in addition to being a naturally detoxifying food) plays beautifully with this briny, herbal dressing. If you want to round it out even further, try tossing a little cooked shrimp or chicken into the mix.

Serves 4

1 small head broccoli, including stems, cut into bite-sized pieces

1 large red bell pepper, cut into bite-sized pieces

1 large green bell pepper, cut into bite-sized pieces

1 large English cucumber, peeled and roughly chopped

1 medium-sized red onion, roughly chopped

2 Roma tomatoes, peeled and roughly chopped

1 cup kalamata olives, pitted and roughly chopped, plus 2 tablespoons olive brine

¼ cup fresh parsley, roughly chopped

⅓ cup red wine vinegar

⅓ cup extra-virgin olive oil

2 tablespoons dried oregano

1 teaspoon ground black pepper

1 teaspoon salt

In a large bowl, toss together the broccoli, bell peppers, cucumber, onion, and tomatoes. Add the olives and parsley and gently fold to combine.

In a medium bowl, whisk together the vinegar, oil, olive brine, oregano, black pepper, and salt. Drizzle the dressing over the salad, carefully toss (so as not to crush the olives), and serve.

Kelp Noodles with Brazil Nut Pesto

I'm a huge fan of pesto, probably because my mom was, too. (She'd also seize any opportunity to link the word *pesto* to *pest*, which she'd lovingly call me.) For that reason, you'll find a few pesto variations in this book. This version is not just utterly delicious, it brings together some powerful, health-boosting ingredients. Uniting Brazil nuts, which are loaded with selenium, with kelp noodles, which are packed with iodine, you have a one-two punch for feeling and looking your best. Plus, this is a keto-friendly "pasta" dish using only healthy fats—what's not to love?

Serves 2 to 3

12 ounces kelp noodles
½ medium lemon
1 teaspoon salt, divided
½ cup extra-virgin olive oil
½ cup fresh basil leaves
1 cup Brazil nuts*

1 teaspoon garlic powder
2 tablespoons nutritional yeast

*Alternatively, you can use ½ cup Brazil nuts and ½ cup macadamia nuts.

Place the kelp noodles in a bowl filled with water and squeeze the lemon half into the bowl, saving 1 teaspoon lemon juice for the pesto. Add ½ teaspoon of the salt, stir lightly, and set aside for 30 minutes.

Combine the oil, basil, Brazil nuts, remaining ½ teaspoon salt, the garlic powder, nutritional yeast, and reserved 1 teaspoon lemon juice in a blender and pulse until coarsely chopped. Set aside.

Drain the kelp noodles and pat dry with a paper towel. Toss the kelp noodles with the pesto and serve.

Roasted Broccoli with Shallots and Dried Fish

Dried fish? Really? Well, even if you're not a fish lover, you'll probably love this dish—the fish adds a bit of saltiness, which plays nicely off the sweetness of the dates. If you're really fish averse (or vegetarian), try using ¼ cup nutritional yeast instead.

Serves 4

4 cups bite-sized broccoli florets (1 large head)
6 shallots, thinly sliced
¼ cup minced dates
¼ cup extra-virgin olive oil

½ cup bonito flakes (available in Japanese supermarkets), minced dried sardines (available in Korean markets), or dried shrimp (found in the international aisle of most well-stocked grocery stores)
2 tablespoons balsamic vinegar

Preheat the oven to 375°F.

In a large bowl, toss together the broccoli, shallots, dates, and oil. Spread on a sheet pan and bake for 15 to 20 minutes, until the broccoli is tender and crisp at the edges.

Remove from the oven and carefully toss with the dried fish of your choice and the vinegar. Return to the oven and bake for an additional 5 to 10 minutes, until the vinegar is syrupy. Serve immediately.

Mustard and Miso Broccoli

The combination of mustard and miso in this dish creates a cheese-like sauce for the broccoli—a great thing for all you broccoli and cheddar lovers out there—but without the dairy. You can omit the walnuts, but I absolutely love the crunch they deliver.

Serves 4

4 cups broccoli florets (1 large head)
3 shallots, roughly chopped
1 cup walnut halves
3 tablespoons walnut oil
1 tablespoon Dijon mustard

1½ teaspoons coconut aminos
2 tablespoons red or white miso paste
Zest and juice of 1 lemon
1 teaspoon monk fruit sweetener
1 clove garlic, minced

Preheat the oven to 400°F.

In a large bowl, toss together the broccoli, shallots, and walnuts.

In a small bowl, whisk together the walnut oil, mustard, coconut aminos, miso, lemon zest and juice, monk fruit, and garlic. Toss the dressing with the broccoli mixture.

Spread out on a sheet pan and bake for 20 to 25 minutes, turning the broccoli a couple times, until the broccoli is crisp and lightly charred. Remove from the oven and serve immediately.

Simple Bone Broth

Bone broth is delicious, easy to make, and highly versatile. It's a staple dish that humans have been likely making for millennia. It also cuts down on food waste since you can use bones left over from dinner. Plus, it's majorly nourishing, providing a bevy of nutrients: collagen, protein, and electrolytes, to name a few. This is a recipe from my friend Amanda Meixner, aka @MeowMeix on Instagram. It utilizes an Instant Pot, which will shorten the cook time, but I've provided alternate, stove-top instructions as well.

Serves 4

Chicken bones from a whole chicken (turkey or beef bones can also be used)

Assorted veggies for flavor (i.e., half an onion, a few carrots and/or celery stalks)

Fresh herbs (thyme, sage, and rosemary are my go-tos)

1 teaspoon Himalayan sea salt

1 tablespoon apple cider vinegar

Water (enough to completely submerge the bones)

Place the bones in Instant Pot and top with veggies, herbs, and salt.

Add apple cider vinegar and then enough water to completely cover the bones and veggies.

Wait an hour before starting the Instant Pot. This will allow the vinegar to pull minerals out of the bones.

Hit the soup button and set 120 minutes on the Instant Pot on the high-pressure setting.

Wait 15 minutes before depressurizing the pot.

Strain the broth and pour into a mug for a simple drink.

Note: If you are making bone broth on a stove, bring water to a boil and simmer for 24 to 72 hours.

Spiced Chickpea and Broccoli Stew with Turmeric and Mint

This stew is inspired by a soup I frequently crave—the restaurant that serves it uses only dry spices, but the addition of fresh turmeric adds a sweet earthiness (and bonus health kick) that plays really well with the nutty chickpeas. Try adding cilantro to the finished product for even more of a fresh zing.

Serves 4

¼ cup extra-virgin olive oil

2 medium yellow onions, finely diced

2 ribs celery, diced

4 cups bite-sized broccoli florets (1 large head)

2 (15-ounce) cans chickpeas, drained and rinsed*

2 tablespoons minced fresh ginger

2 tablespoons minced fresh turmeric**

1 teaspoon ground cumin

1 teaspoon ground black pepper

¼ teaspoon ground cinnamon

2 cups chicken bone broth or vegetable broth

1 cup plain unsweetened coconut yogurt, divided

½ cup sliced fresh mint

*You can cook your own chickpeas (I suggest pressure cooking), but using high-quality canned is a great time-saver on a busy night. If desired, substitute 3 cups cooked chickpeas for 2 cans.

**If you can't find fresh turmeric, use 1½ teaspoons dried turmeric instead.

Heat the oil in a large soup pot over medium-high heat. Add the onions and celery and cook, stirring occasionally, for 4 to 6 minutes, until tender.

Add the broccoli, turn the heat up to high, and cook for 3 to 5 minutes to get a char on the broccoli. Add the chickpeas, ginger, and fresh turmeric (if using dried instead, wait until the next step) and cook until very fragrant, 1 to 2 minutes.

Add the cumin, black pepper, cinnamon, and dried turmeric, if using, and cook for 1 additional minute.

Add the broth, bring to a simmer, then reduce the heat to low and cook until the stew is nice and thick, 20 to 25 minutes. Fold half of the yogurt and all of the mint into the stew. Serve, with dollops of the remaining yogurt on top.

Broccoli "Falafel" with Creamy Avocado Sauce

Growing up in New York City, I had access to the best falafel outside of the Middle East! Unfortunately, most are deep-fried in unhealthy oils and also not the best option if you're trying to limit starch in your diet. Not too long ago, I wondered, *Could baked falafel ever be as delicious as fried?* My imagination ran wild, and the next thing you know, I was in my kitchen experimenting with the vegetables I had on hand. I discovered that the combination of broccoli and walnuts makes a beautiful—and less starchy—falafel, which goes perfectly with the creamy avocado-sesame sauce in this recipe.

Serves 4

2 cups broccoli florets (½ large head)
1 medium red onion, cut into small pieces
1 cup raw walnuts
1 teaspoon salt, divided
4 tablespoons avocado oil, divided
Zest of 3 lemons, divided
½ teaspoon paprika
½ teaspoon ground cumin
½ teaspoon dried oregano
¼ cup toasted sesame seeds

½ cup fresh parsley leaves
¼ cup ground flaxseeds, plus more as needed
¼ cup tahini (sesame seed paste)
1 tablespoon sesame oil
1 avocado, flesh scooped out
Juice of 2 lemons
1 clove garlic, peeled
¼ cup fresh dill, roughly chopped

Preheat the oven to 400°F.

In a large bowl, combine the broccoli, onion, walnuts, and a ½ teaspoon of the salt. Add 2 tablespoons of the avocado oil and toss to coat the ingredients.

Transfer to sheet pan and roast for 15 to 20 minutes, until the broccoli is tender and slightly charred. Remove from oven and let cool until easy to handle. Do *not* turn off the oven.

Transfer the broccoli mix to a food processor fitted with an "S" blade and add the zest of 1 lemon, the paprika, cumin, oregano, sesame seeds, parsley, and flaxseeds. Pulse until well combined, then let rest for 15 minutes.

If the mixture is too wet to shape, add more flaxseeds. If it's too dry or crumbly, add water, 1 teaspoon at a time. With damp hands, shape the mixture into small 2-bite balls.

Heat the remaining 2 tablespoons avocado oil in a large skillet over medium-high heat. Add the falafel (you'll likely need to work in batches) and sear for 1 to 2 minutes per side, until golden brown. Transfer to a sheet pan and bake for 5 to 10 minutes to warm through.

While the falafel is in the oven, make the sauce: Combine the remaining salt, the tahini, sesame oil, avocado, remaining lemon zest, the lemon juice, garlic, and dill in a blender or food processor fitted with an "S" blade. Blend until creamy, thinning with water as needed. Serve with the falafel.

Broccoli Walnut Bisque

This is one of the most indulgent recipes in this book—it's creamy, rich, and nutty, and it truly feels like a special occasion food. The key to making this soup taste rich and nutty, rather than a bit sulfuric (as broccoli sometimes does) is to thoroughly cook the vegetables to really develop their flavor before simmering the soup.

Serves 4 to 6

4 tablespoons walnut oil, divided
2 shallots, minced
1 clove garlic, minced
4 cups broccoli florets (1 large head)
6 cups chicken bone broth or vegetable broth

2 cups toasted walnuts, soaked in 2 cups water for 8 hours or up to overnight
Juice of 1 lemon
¼ cup roughly chopped toasted walnuts
Unsweetened coconut cream

Heat 2 tablespoons of the oil in a large soup pot over medium-high heat. Add the shallots, garlic, and broccoli and cook until the shallots and garlic are tender and fragrant and the broccoli is beginning to crisp up. Reduce the heat to medium and add the broth, soaked walnuts and their soaking liquid, and lemon juice to pot. Bring to a simmer, then reduce the heat to maintain a simmer, cover, and cook until the broccoli is just tender, about 8 minutes.

With a hand blender or in a standing blender in batches, blend the soup until smooth and creamy (I like to use a high-powered blender to really get the walnuts broken down).

Return the soup to the pot, place over medium heat, and simmer to thicken or stir in additional broth to thin to your desired consistency (I like mine very thick and creamy).

Serve drizzled with the remaining 2 tablespoons oil, the toasted walnuts, and a small dollop of coconut cream.

Broccoli "Latkes" with Blueberry Applesauce

Come the holiday season, what could you always guarantee was cooking in the Lugavere household? Latkes. I don't care what anyone says, my grandma Hilda made the best in Murray Hill! These, well . . . these are pretty far from traditional latkes. And you know what? That's okay. When you combine these crispy, salty broccoli fritter/pancakes with a tangy, sweet sauce, the result is magical, whether you're celebrating a holiday or just making a weeknight dinner.

Serves 4 to 6

3 large apples, peeled and roughly chopped*
1 cup fresh or frozen blueberries
¼ cup water
2 tablespoons monk fruit sweetener
1 cinnamon stick
Zest and juice of 1 lemon
1 small yellow onion, roughly chopped
1 medium head broccoli, roughly chopped
2 eggs

½ teaspoon salt
3 tablespoons almond flour
2 tablespoons tapioca starch, plus more as needed
3 tablespoons avocado oil, plus more as needed

*I like Gala, Golden Delicious, or Fuji for this recipe, but any sweet apple will work.

In a large soup pot with a tight-fitting lid, combine the apples, blueberries, water, monk fruit, cinnamon, and lemon zest and lemon juice. Place over low heat, cover, and cook for 20 to 30 minutes, until the apples fall apart. Using a potato masher, mash the apples for a chunky sauce (blend for a smoother sauce) and continue to cook until excess water evaporates. Remove from the heat and set aside.

While applesauce is cooking, make the latkes: In a food processor fitted with an "S" blade, pulse the onion and broccoli until broken down to the size of grains of rice. You can also grate the onion and broccoli on the large holes of a box grater.

Transfer to a large bowl and add the eggs, salt, almond flour, and tapioca starch.

Heat the oil in a large skillet or on a griddle over medium heat. Add the broccoli mixture in 3-tablespoon scoops (or smaller for bite-sized latkes). Cook for 2 to 3 minutes on each side, until golden brown and crisp. Serve immediately, topped with the applesauce.

Fish Sauce Caramel–Glazed Beef Spring Rolls

These lettuce-wrapped spring rolls are inspired by traditional Vietnamese spring rolls, but lightened up by removing the rice paper wrappers. It takes a bit of practice to wrap these rolls perfectly, but once you get it, it's a skill you'll have forever—and if you don't . . . you still win, because you end up with a delicious salad instead. Try to make with a high-quality fish sauce, which is usually just fish and salt.

Serves 4

¼ cup fish sauce
2½ tablespoons monk fruit sweetener, divided
2 tablespoons fresh lime juice
2 cloves garlic, minced
1 pound flank or skirt steak
2 tablespoons sesame oil
8 to 12 Bibb lettuce leaves

2 medium carrots, cut into matchsticks
1 jalapeño, cut into matchsticks
1 cup bean sprouts
¼ cup fresh mint leaves
¼ cup fresh Thai basil leaves
¼ cup fresh cilantro leaves
¼ cup fresh shiso leaves (optional but nice)

For the dipping sauce:
¼ cup unsweetened almond butter
2 tablespoons coconut aminos or tamari
1 tablespoon fish sauce

Juice of 1 lime
1 teaspoon grated fresh ginger

In a small saucepan, combine the fish sauce, 2 tablespoons of the monk fruit, the lime juice, and garlic and heat over medium-low heat until the monk fruit is dissolved. Let cool to room temperature, then pour over the steak. Marinate in refrigerator for at least 3 hours or up to overnight.

When ready to cook, remove the meat from the marinade, reserving the marinade. Pat the steak dry.

Heat a large skillet over high heat and add the sesame oil. Add the steak and sear for 3 to 4 minutes on each side, until brown and caramelized on the outside, then remove from the heat to a cutting board. Add the remaining marinade to skillet, along with the remaining ½ tablespoon monk fruit and all the dipping sauce ingredients and cook, stirring occasionally, until the monk fruit is dissolved and the mixture comes together. Let cool.

Slice the steak into thin, easy-to-bite pieces and place 1 or 2 pieces in the center of each lettuce leaf. Add the carrots, jalapeño, bean sprouts, mint, Thai basil, cilantro, and shiso, if using, and roll like you would a burrito. Serve with the dipping sauce on the side.

Note: Instead of making spring rolls, you could chop the steak, lettuce, bean sprouts, herbs, and pickled vegetables into a salad and use the dipping sauce as a salad dressing.

Blackberry, Avocado, and Basil Salad

Blackberries in a salad? You're damn right! You'll be impressed with how nicely the tangy blackberries play off the creaminess of the avocado. If you're not a coconut fan, you can absolutely leave it off to the side—but I seriously suggest trying it. It adds a toasty flavor rather than an overwhelming coconut flavor.

Serves 4

2 large shallots, minced

Juice of 2 limes

2 cups blackberries, sliced in half

1 cup toasted pecan halves

½ cup unsweetened shredded coconut, toasted

5 cups baby spinach, loosely packed

2 cups fresh basil leaves, roughly torn

2 avocados, cubed

¼ cup extra-virgin olive oil

¼ cup red wine vinegar

In a small bowl, cover the shallots in the lime juice and set aside for 15 to 20 minutes to tame the bite of the shallots.

In a large bowl, toss together the blackberries, pecan halves, coconut, spinach, basil, and avocados.

Add the oil and vinegar to the shallot and lime juice mixture, pour the dressing over the salad, toss, and serve.

Beef Meatball, Garlic, and Greens Soup

New York City became a melting pot in the early twentieth century, attracting people from all over the world who sought a better life for themselves and their families. That's probably why Italian food is almost as familiar to my palate as traditional Jewish food. This dish is a riff on Italian wedding soup, featuring mini meatballs (or big meatballs if you'd prefer), hearty greens, and plenty of garlic to punch up the flavor. This is sort of a three-for-one recipe deal; you can also use the meatballs to go with sweet potato (or other grain-free) noodles, or to use as a topping on a cauliflower crust pizza.

Serves 4 to 6

1 pound ground beef
1 teaspoon garlic powder
1 teaspoon onion powder
1 teaspoon paprika
1 teaspoon ground black pepper
½ teaspoon salt, or to taste
1 egg, beaten

1 tablespoon tapioca starch, plus more as needed
¼ cup extra-virgin olive oil
1 head garlic, cloves thinly sliced
6 cups beef or chicken bone broth
1 bunch lacinato kale, ribs removed, leaves thinly sliced
1 bunch Swiss chard, thinly sliced
Hot sauce (optional)

First, make the meatballs: In a large bowl, combine the beef, garlic powder, onion powder, paprika, black pepper, salt, egg, and tapioca starch. Fold together with clean hands until well mixed and check that the mixture forms into a ball easily—if it's overly wet, add tapioca starch 1 teaspoon at a time.

Fill a small bowl with water and wet your hands (it's easier to work the meat with wet hands). Shape the mixture into bite-sized meatballs, place on a plate or tray, and refrigerate for at least 30 minutes.

Heat the oil in a large soup pot over medium-high heat. Add the meatballs and sear on all sides, about 1 minute per side, until golden brown and lightly crusted on the outside. Add the garlic to the pot and cook until the garlic is very fragrant and golden brown. Add the broth, bring to a simmer, then lower the heat, cover, and simmer for 20 minutes for the meatballs to cook through and to infuse the soup with garlic flavor. Add the sliced greens and cook, uncovered, stirring occasionally, until the greens are wilted, about 8 minutes. Taste and season with salt and/or hot sauce and serve.

Spicy Sesame Broccoli Slaw

What do New York Jews always eat at Christmas time? Chinese food. (The phenomenon is so ubiquitous and dates back so long—over one hundred years, to be precise—it has even been studied in the academic literature.)[46] This slaw reminds me a bit of a classic Chinese chicken salad, which, let's admit, is probably an American creation. Nonetheless, what we have here is even better—sweet, tangy, nutty, creamy, and nicely crisp. It pairs well with anything barbecued or with a simple seared chicken or salmon dish.

Serves 4

3 cups broccoli slaw*
1 medium red onion, thinly sliced
1 jalapeño, thinly sliced
1 cup shredded red cabbage
½ cup fresh cilantro leaves
1 avocado, chopped
¼ cup tahini (sesame paste)
1 tablespoon toasted sesame oil
1 teaspoon rice vinegar
Juice of 1 lemon

1 teaspoon monk fruit sweetener (optional)
1 to 3 teaspoons coconut aminos or tamari
2 ripe oranges, segmented (optional)
Sriracha
3 tablespoons toasted sesame seeds

*Available in the prepackaged veggie selection in most well-stocked grocery stores. To make it yourself, it's a mixture of shredded broccoli stems, cabbage, and carrots.

In a large bowl, toss together the slaw, onion, jalapeño, cabbage, and cilantro until well combined.

In a blender or a food processor fitted with an "S" blade, combine the avocado, tahini, oil, vinegar, lemon juice, monk fruit, and 1½ teaspoons coconut aminos and blend until very smooth and creamy.

Gently toss the dressing and oranges into the slaw mixture. Taste and add sriracha and coconut aminos as you like. Serve topped with the toasted sesame seeds.

Garlic and Curry Snacking Almonds

This dish looks so simple in the bowl—just seasoned almonds. But once you get a bite, you'll be hooked—the spices are rounded out by naturally sweet dried coconut for an unbeatable flavor. I keep these on hand just about all the time—and rely on them as a topping for a simple salad or a protein-rich snack to hold me over in between meals.

Serves 4 to 6

1 tablespoon extra-virgin olive oil
2 cups lightly toasted almonds
1 egg white, beaten with a fork until foamy
2 cloves garlic, grated on a Microplane

1 tablespoon curry powder
½ teaspoon ground cumin
1 teaspoon salt
¼ cup unsweetened shredded coconut

Preheat the oven to 325°F. Line a sheet pan with parchment paper and drizzle with the oil.

In a large bowl, toss the almonds with the egg white, making sure each almond is coated. Add the garlic and toss again to combine.

In a small bowl, combine the curry powder, cumin, salt, and shredded coconut.

Add the spice mixture to the nut mixture and stir well to thoroughly coat. Transfer to the prepared sheet pan and bake for 30 minutes, pausing to stir mixture every 10 minutes. Let cool before enjoying.

Avocado and Ginger Chicken Spring Rolls

Growing up in New York City, I had access to incredible Chinese food. One of my favorite treats was spring rolls, which became off-limits when I learned that they're usually deep-fried in unhealthy oils. Enter this delicious and healthy twist. I love the addition of avocado to these ginger-rich lettuce-wrapped spring rolls. The creaminess balances out the minor bite from the ginger and the heat from the dipping sauce. If you have trouble mastering the art of wrapping these spring rolls, that's okay—chopping the lettuce with the other ingredients makes a great salad!

Makes 8 to 12

3 tablespoons sesame oil
1 shallot, minced
3 cloves garlic, minced
3 tablespoons minced fresh ginger
1 pound ground chicken
2 tablespoons coconut aminos or tamari

Butter lettuce leaves, for wrapping
1 large carrot, cut into matchsticks
¼ cup fresh mint leaves
¼ cup fresh basil leaves
2 avocados, thinly sliced

For the dipping sauce:
1 cup hot water
½ cup monk fruit sweetener
Juice of 2 limes

½ cup fish sauce
2 cloves garlic, minced
1 jalapeño or bird's eye chile, minced

Heat the oil in a large skillet over medium-high heat. Add the shallot, garlic, and ginger and cook for 2 to 3 minutes, until very fragrant. Add the chicken and cook until the chicken is browned and cooked through, about 7 to 10 minutes. Add the coconut aminos and simmer down until the liquid evaporates. Remove from the heat and let the mixture cool to room temperature.

To make the dipping sauce, in a small bowl, combine the hot water and monk fruit and stir until the monk fruit is dissolved. Add the lime juice, fish sauce, garlic, and chile and stir to combine.

Assemble your spring rolls: Place a couple spoonfuls of the chicken mixture in the center of a lettuce leaf and follow it with some carrot slices, a few mint and basil leaves, and some avocado slices. Roll like you would a burrito, dip, and eat.

Avocado, Fennel, Pomegranate, and Winter Citrus Salad

This is one of my favorite salads of all time. I'm such a big fan of raw fennel that I almost decided to name my podcast *The Max Lugavere Raw Fennel Love Show*. It's a riff on a classic citrus and fennel salad featuring fruits and veggies that are in season in the winter, along with hearty, peppery arugula to balance out the sweetness of the citrus and apples.

Serves 4

1 shallot, minced
¼ cup red wine vinegar
Zest and juice of ½ lemon
¼ cup extra-virgin olive oil
1 teaspoon Dijon mustard
2 cups arugula
1 bulb fennel, cored and thinly sliced
1 green apple, cored and cut into matchsticks

1 grapefruit, peel removed, cut into half-moons
2 blood oranges, peel removed, cut into rounds
2 tangerines, peel removed, cut into rounds
¼ cup kumquats, thinly sliced
2 avocados, thinly sliced
½ cup pomegranate arils
½ cup minced fresh mint

In a large bowl, whisk together the shallot, vinegar, lemon zest and juice, the oil, and mustard until well combined.

In a large bowl, toss together arugula, fennel, and apple and spread onto a serving platter. Top with the sliced grapefruit, blood oranges, tangerines, and kumquats, spreading them around for even distribution.

Spread the avocados on top of the citrus and immediately drizzle with the dressing to prevent the avocado from browning. Top with the pomegranate and mint and serve.

Zucchini and Carrot Noodles with Herbed Avocado "Pesto"

As someone who tends to avoid typical grain-based pastas, I appreciate that more and more stores are carrying vegetable "noodles" in the precut veggie section—and they're also super easy to make on your own. I find that the ultra-creamy pesto in this is the perfect companion to the crisp veggie noodles—and if you want a real treat, consider pairing this with Olive Oil–Poached Salmon (page 228).

Serves 4

3 cups zucchini noodles (about 3 large zucchini, spiralized or cut into thin, spaghetti-sized strips)

2 cups carrot noodles (about 3 large carrots, spiralized or cut into thin, spaghetti-sized strips)

1½ teaspoons salt, divided

4 tablespoons extra-virgin olive oil, divided

1 teaspoon ground black pepper

½ teaspoon red chile flakes

1 large avocado

1 clove garlic, peeled

Zest and juice of 1 lemon

¼ cup minced fresh dill

¼ cup minced fresh parsley

1 cup fresh basil leaves

¼ cup nutritional yeast

½ cup toasted slivered almonds

Place the zucchini and carrot noodles in a large bowl and sprinkle with ½ teaspoon of the salt. Set aside for 5 to 10 minutes to draw out moisture. Pat the noodles dry.

Heat 2 tablespoons of the oil in a large skillet over medium-high heat. Add the noodles, ¼ teaspoon of the salt, black pepper, and chile flakes, and cook for 5 to 7 minutes, stirring frequently, until the noodles are tender and crisp at the edges. Allow to cool slightly while you make the sauce.

In a food processor fitted with an "S" blade, combine the avocado, garlic, lemon zest, dill, and parsley and process until smooth. Add the basil and nutritional yeast and pulse to combine.

With the motor running, stream in the lemon juice and remaining 2 tablespoons oil until a smooth, creamy sauce is formed. Taste and season with the remaining salt, ¼ tsp at a time, until your taste buds are satisfied.

Off the heat, toss the noodles with the sauce, top with the slivered almonds, and serve.

"Cheesy" Kale Salad

Here's a delicious salad that's easy to make and savory enough to convert even the most salad-phobic of the bunch. If you really want to go crazy (live a little!), find some anchovies packed in either extra-virgin olive oil or water at your local market, chop them lightly, and toss with the dressing. Though this salad is perfect as a starter, I love to throw a Simple Burger Patty (see page 212) on top of it to make a complete meal.

Serves 2 to 3

1 bunch kale, center ribs and stems removed
2 tablespoons extra-virgin olive oil
2 tablespoons apple cider vinegar
1/2 large green bell pepper, chopped

1/4 cup nutritional yeast
1 teaspoon garlic powder
3/4 teaspoon salt

Tear the kale leaves into small pieces and place them in a large bowl. Add the oil and vinegar and stir or massage it into the leaves to start to soften them. Add the green pepper, then the nutritional yeast, garlic powder, and salt and toss until everything is well combined.

Charred Eggplant with Chiles and Avocado

Charring the eggplant in this dish adds a slightly sweet, really nutty flavor, one that balances out the spiced extra-virgin olive oil the dish is cooked in. Plus, the addition of sweet pomegranates, fresh mint, and creamy avocado is the perfect balance for the rich, spicy eggplant that's the star of this dish.

Serves 4

2 large eggplants, cut into ½-inch rounds

1½ teaspoons salt, divided

4 tablespoons extra-virgin olive oil, divided

1 large red onion, cut into rings

2 Fresno or other hot chiles, seeds removed, cut into rings

½ teaspoon ground cumin

½ teaspoon ground black pepper

2 avocados, skin and pit removed

Zest and juice of 1 lemon

¼ cup minced fresh mint

½ cup pomegranate arils

Sprinkle the eggplant rounds with ¾ teaspoon of the salt and let rest for 15 to 20 minutes.

In a large saute pan, heat 3 tablespoons of the oil over medium-low heat. Add the onion and chile, reduce the heat to low, and cook for 10 to 15 minutes, stirring occasionally, until the onion is very tender. Remove the vegetables from the pan to a bowl using a slotted spoon and let the oil cool to room temperature.

Carefully transfer the oil to a heatproof measuring cup or other container that's easy to pour from.

Pat the eggplant dry with a kitchen towel, using pressure to squeeze out as much of the excess liquid as possible.

Heat 1 tablespoon of the chile-infused oil in the skillet over high heat. Working in batches, cook the eggplant rounds, being careful not to crowd the pan, for 3 to 5 minutes per side, until charred all over, replenishing the oil as needed.

Arrange the eggplant on a serving tray and carefully pat with paper towels to remove excess oil. Top with the onion and chile. Combine the cumin, the remaining ¾ teaspoon salt, and the black pepper and sprinkle the spice mixture on top.

Whisk together the remaining oil, ripe avocados, lemon zest, and lemon juice until well combined. Top the eggplants with dollops of the avocado mixture, sprinkle with the mint and pomegranate arils, and serve.

Broccoli "Rice" with Pomegranate and Fresh Herbs

Just like cauliflower rice, broccoli rice is making quite the splash in many produce sections, and with good reason—it's a great alternative to traditional rice. And when tossed with plenty of fresh herbs, salty pistachios, and sweet pomegranate, this dish is more than just a healthy take on rice—it's flavorful, delicious, and totally crave-able.

Serves 4

4 tablespoons extra-virgin olive oil, divided
1 medium yellow onion, minced
4 cups broccoli rice*
1 teaspoon ground turmeric
1 teaspoon ground black pepper
1 teaspoon salt
1/2 cup minced fresh parsley
1/2 cup minced fresh cilantro
1/4 cup minced fresh dill

1/4 cup minced fresh mint
1 1/2 cups pomegranate arils
3/4 cup salted shelled pistachios

*Available in the packaged veggie section of most major grocery stores. To make your own, pulse stems of broccoli in a food processor to the size of a grain of rice.

In a large saute pan, heat 2 tablespoons of the oil over medium-high heat. Add the onion and cook for 2 to 3 minutes, until translucent. Add the broccoli rice, turmeric, black pepper, and salt and cook for 7 to 10 minutes, until the broccoli rice is tender. Remove from the heat and fold in the parsley, cilantro, dill, mint, and pomegranate arils. Top with the pistachios and serve.

Lemongrass, Mint, and Basil Roasted Broccoli

I'm always puzzled by people who don't enjoy broccoli. I wonder, *Have you just never had it properly prepared?* For even the staunchest of broccoli haters, this dish is really special. The combination of sesame oil, lemongrass, and mint offset the nuttiness of the roasted broccoli and sesame seeds beautifully, and basil adds a bright herbal note that reminds me of Vietnamese food—one of my favorites.

Serves 4

2 tablespoons avocado oil

1 tablespoon sesame oil

4 tablespoons very finely chopped lemongrass, divided

4 cups broccoli florets (1 large head)

¼ cup minced fresh mint

¼ cup minced fresh basil

¼ cup toasted sesame seeds

1 to 2 teaspoons coconut aminos or tamari

Preheat the oven to 425°F.

In a small saucepan, combine the avocado oil, sesame oil, and 2 tablespoons of the lemongrass. Cook over low heat for 5 to 10 minutes to infuse the lemongrass flavor into the oil.

Toss the oil with the broccoli and remaining 2 tablespoons lemongrass. Spread on a sheetpan and bake for 10 to 15 minutes, until the broccoli is tender and crisp at the edges. Toss with the mint and return to the oven for 2 to 3 minutes. Remove from the oven, toss with the basil, sesame seeds, and coconut aminos, and serve.

Crispy Broccoli Hot "Wings"

I'll be the first to admit—traditional chicken hot wings are delicious, and you can find my recipe for the best (and, shockingly, best for you) you've ever had on page 194. But there's no reason you can't love these broccoli "wings" equally as much, especially as a side dish. Plus, they're the perfect crowd pleaser for a game day or party. They've got the same crispy, spicy kick as buffalo wings, too.

Serves 6 to 8

3 tablespoons avocado oil, plus more for the pan

1 cup unsweetened canned coconut or almond milk

1 cup plus 1 tablespoon cayenne hot sauce (Tabasco, Crystal, or Frank's RedHot preferred), divided

½ cup tapioca starch

1½ cups almond flour, divided, plus more as needed

1¼ cups ground flaxseeds, divided, plus more as needed

1 teaspoon salt

4 cups broccoli florets (1 large head)

4 cloves garlic, minced

¼ cup plain unsweetened coconut yogurt

½ teaspoon fresh lemon juice

1 teaspoon apple cider vinegar

¼ teaspoon celery salt

¼ teaspoon ground black pepper

2 tablespoons nutritional yeast

Preheat the oven to 425°F. Lightly grease a sheet pan with oil and place inside the oven to preheat the pan (this will give you crispier "wings").

In a shallow bowl, whisk together the milk, 1 tablespoon of the hot sauce, the tapioca starch, ½ cup of the almond flour, ¼ cup of the flaxseeds, and the salt to make a batter.

In a separate bowl, whisk together the remaining 1 cup almond flour and 1 cup flaxseeds.

Dip the broccoli florets in the batter, then transfer to the dry mixture and toss to coat well. If you run out of dry mixture, replenish with equal parts almond flour and flaxseeds.

When all the broccoli is coated, transfer to the hot sheet pan (carefully—don't burn yourself) in a single layer. Bake for 8 minutes, then flip and bake for another 5 to 7 minutes, until nicely browned.

Meanwhile, heat the oil and garlic in a small saucepan over medium heat for 2 to 3 minutes, until the garlic is softened and fragrant. Add the remaining 1 cup hot sauce and remove from the heat.

To make the dipping sauce: Whisk together the coconut yogurt, lemon juice, vinegar, celery salt, black pepper, and nutritional yeast.

When the wings come out of the oven, gently drizzle with the hot sauce mixture (or go for a full dip if you're feeling bold). Serve immediately.

Junk-Free, Dairy-Free Spinach and Artichoke Dip

There was a time in my life when the dip was the first thing I'd go for on a party buffet table, and I'm happy to say that when it comes to this dip, loaded with some of my favs like nutritional yeast and extra-virgin olive oil, I'm right there again. It's a healthy way to enjoy the party classic—just make sure to go for healthy dippers such as sweet potato chips or, even better, raw veggies!

Serves 4

2 cups raw cashews, macadamia nuts, or peeled almonds, soaked in water overnight and drained

¾ cup water

Zest and juice of 1 lemon

1 teaspoon apple cider vinegar

1 teaspoon paprika

1 teaspoon onion powder

1 teaspoon salt, or to taste

¼ cup nutritional yeast, or to taste

1 tablespoon extra-virgin olive oil

3 cups chopped baby spinach

3 cloves garlic, chopped

1 (14-ounce) can quartered artichoke hearts, drained and roughly chopped

Splash of unsweetened canned coconut milk (optional)

Make your "cheese" sauce: Combine the cashews, water, lemon zest and juice, vinegar, paprika, onion powder, salt, and nutritional yeast in a food processor fitted with an "S" blade and process to a very smooth, creamy sauce, at least 3 to 4 minutes.

Heat the oil in a large skillet over medium-high heat. Add the spinach and garlic and cook until the garlic is fragrant and the spinach is wilted, about 4 minutes. Add the artichoke hearts and cook to release extra moisture, 2 to 3 minutes, then transfer the spinach mixture to a bowl.

Fold in the "cheese" sauce and taste—add salt if you'd like it saltier, more nutritional yeast for a "cheesier" flavor, or a splash of coconut milk to make it creamier. Serve with sweet potato chips or veggies for dipping.

Garlicky Chard with Dried Blueberries

Greens sauteed with garlic is one of my all-time favorite simple side dishes—but when I want to take it up a notch, there's nothing like toasted nuts and tangy dried blueberries to round out the flavor and elevate the dish. The nuttiness and sweetness play nicely off the bite of the garlic, for a "party in your mouth" with every bite.

Serves 4

¼ cup extra-virgin olive oil
8 cloves garlic, thinly sliced
¼ cup slivered blanched almonds
1 cup unsweetened dried blueberries

2 bunches Swiss chard, stems and leaves
  separated and chopped separately
¼ cup balsamic vinegar
½ teaspoon salt
½ teaspoon ground black pepper

Heat the oil in a very large skillet over medium heat. Add the garlic and almonds and cook, stirring frequently, until the ingredients are golden brown and the garlic is fragrant, 2 to 3 minutes. Using a slotted spoon, scoop the garlic and almonds from the oil and set aside for later.

Increase the heat to medium-high, add the blueberries and chard stems to the pan, and cook until the chard stems are tender, about 7 minutes. Add the chard leaves and cook, tossing frequently, until wilted and tender, about 5 minutes. Add the vinegar, salt, and black pepper, increase the heat to high, and cook, stirring frequently, until the vinegar is reduced to a syrupy consistency. Remove from the heat, return the garlic and almonds to the pan, mix them in, and serve.

Israeli-Style Pickled Cabbage

I looked into my ancestry with a popular gene-testing site and found out that I'm 99.8 percent Ashkenazi Jew. That means I'm hardwired to enjoy two things: Larry David and anything pickled. Hamutzim are Israeli-style pickled veggies that are often served before a big meal to perk up your taste buds and get your mouth watering. This simple quick-pickled cabbage is great paired with something creamy like hummus or avocado, or with rich grilled beef, and it's also great in place of traditional slaw.

Serves 6 to 8

1 small head red cabbage
4 tablespoons salt, divided
1 cup red wine vinegar
1½ cups distilled white vinegar
¼ cup monk fruit sweetener
Zest and juice of 1 lemon (cut the zest into large strips)

¼ teaspoon dill seeds
2 cloves garlic, peeled
1 teaspoon mustard seeds
1 small cinnamon stick
1 serrano pepper, halved
1 bunch dill

Loosely chop the cabbage and spread it on a sheet pan. Sprinkle with 2 tablespoons of the salt and let sit for at least 2 hours or up to overnight (counter is fine or fridge if you live in a hot climate).

When ready to pickle, combine the red wine vinegar, distilled vinegar, the remaining 2 tablespoons salt, the monk fruit, and lemon zest and juice in a medium saucepan and heat over medium low heat until the salt and sweetener have dissolved.

Pat the cabbage dry (be careful, it will stain), then tightly pack it into jars, dividing the dill seeds, garlic, mustard seeds, cinnamon, hot pepper, and dill evenly among the jars (you will need to break up the cinnamon stick unless you are using one very big jar).

Divide the brine equally among the jars. It should cover the cabbage—if not, add a bit more distilled vinegar to cover.

Let sit covered at room temperature for at least 4 hours. The longer it sits, the more pungent it will be. Store the pickled cabbage in the refrigerator, where it will keep for 2 to 3 weeks.

Kale, Arugula, and Parsley Tabouli Salad

Tabouli without the cracked wheat . . . why not? Conventions were made to be broken. This fresh, herbal salad is one of the easiest recipes in this book—all you need to do is chop and toss and you're good to go! Just don't skimp on the lemon, which gives this salad its signature punch.

Serves 4

4 to 5 cups fresh curly parsley, minced
2 cups kale leaves, minced
1 cup arugula, minced
2 cups cauliflower rice
1/3 cup hulled hemp hearts
1 medium red onion, minced

2 ripe Roma tomatoes, minced
1 medium English cucumber, minced
Juice of 2 lemons
1/4 cup extra-virgin olive oil
1/2 teaspoon salt, or to taste

In a large bowl, toss together the parsley, kale, arugula, cauliflower rice, hemp hearts, onion, tomatoes, and cucumber. Add the lemon juice, oil, and salt and toss to coat. Taste and adjust the salt as needed.

Let rest at room temperature for at 30 minutes on the counter or up to overnight in the refrigerator (the flavors really meld after resting). Serve at room temperature.

Kale, Garlic, and White Bean Soup with Orange Zest

I love a garlic-heavy soup in the winter, especially when it's hearty enough to be a stand-alone meal. And thanks to the beans in this soup, it really delivers. Yes, it's a lot of garlic, so if you're making this as a date-night food, make sure you both enjoy it. And don't skimp on the orange zest—it doesn't make the soup sweet but rounds out the flavor nicely. I've included an option to add sausage, which gives this starter main dish potential!

Serves 4 to 6

¼ cup extra-virgin olive oil
¼ cup chopped hard chorizo or other
  sausage (optional)
1 large yellow onion, diced
1 head garlic, cloves diced
2 tablespoons thyme
1 teaspoon salt, or to taste
1 teaspoon ground black pepper, or to taste
½ teaspoon paprika
Zest and juice of 1 lemon
Zest of 1 orange

2 bunches kale, stems removed, leaves
  shredded
2 (15-ounce) cans cannellini or other white
  beans*
8 cups chicken bone broth or vegetable
  broth

*You could soak and cook dried beans, but this is a good time-saver for a busy night. If starting from dried beans, cook thoroughly before proceeding with the recipe (¾ cup dried beans per can).

Heat the oil in a large heavy-bottomed soup pot over medium heat. Add the chorizo, if using, the onion, and garlic and cook until the onion is tender, about 5 minutes, stirring frequently so the garlic doesn't burn. Add the thyme, salt, pepper, paprika, lemon zest, and orange zest and cook for an additional 1 to 2 minutes, until fragrant.

Add the kale and white beans and cook, stirring, until the kale wilts. Add the broth, bring to a simmer, then reduce the heat, cover, and simmer for 25 minutes. Uncover and cook for an additional 5 to 10 minutes, until kale is very tender and beans are beginning to fall apart. Add the lemon juice and serve.

Shrimp and Roasted Veggie Salad

I get bored easily (what can I say? I'm a Gemini). That's why, when constructing a salad, I always make sure there's a variety of chopped veggies that'll keep every bite interesting. This salad is perfectly balanced, thanks to the fresh herbs, tangy, creamy dressing, and ample nutty roasted veggies rounding out each bite. Plus, it's on an arugula base, which means that this entire salad is a nitrate powerhouse. Chew slowly and enjoy this salad's targeted blood pressure benefits.

Serves 4

3 tablespoons extra-virgin olive oil, divided, plus more for the pan
2 tablespoons minced fresh parsley
2 tablespoons minced fresh rosemary
2 cloves garlic, minced
1 teaspoon Dijon mustard
1 pound shrimp, peeled and deveined (tail-on is OK)
1 large red onion, finely chopped
1 bunch asparagus, ends removed, cut into bite-sized pieces

2 zucchinis, diced
2 cups bite-sized broccoli florets (½ large head)
1 cup cherry tomatoes
2 tablespoons red wine vinegar
1 avocado, chopped
Juice of 1 lime
1 jalapeño, seeds removed
¼ cup fresh cilantro, roughly chopped
1 teaspoon salt
4 cups loosely packed arugula

Preheat the oven to 400°F. Lightly oil a sheet pan.

In a large bowl, whisk together 2 tablespoons of the oil, the parsley, rosemary, garlic, and mustard. Add the shrimp, onion, asparagus, zucchinis, broccoli, and cherry tomatoes to the bowl and toss everything until coated in the olive oil mixture.

Pour the shrimp and veggies onto the sheet pan and spread in an even layer. Roast for 15 to 20 minutes, until the shrimp is cooked through and the veggies are tender.

While the shrimp is roasting, make the dressing: In a blender, combine the remaining 1 tablespoon oil, the vinegar, avocado, lime juice, jalapeño, cilantro, and salt and blend until smooth.

Place the arugula on a serving tray and arrange the roasted veggies on top. Drizzle with the avocado dressing and serve.

Olive Oil Mashed Celery Root

I know, I know. Everybody loves mashed potatoes, which—sorry to ruin the party—are usually fat and starch bombs. This mashed celery root mixture (which does includes a sweet potato) is the perfect healthier swap-in. It's flavored with a combination of simmered garlic and toasty garlic oil without added dairy, and it's a super versatile side for meat-based dishes, the garlicky shrimp on page 201, or even with eggs for breakfast.

Serves 4

2 large celery roots, peeled and diced

1 medium white sweet potato, peeled and diced

2 cups chicken bone broth or vegetable broth, plus more as needed

1 head garlic, cloves peeled, divided

1/4 cup extra-virgin olive oil, plus more as needed

Salt and ground white pepper

Place the celery roots, sweet potato, broth, and 5 or 6 cloves of garlic in a large saucepan, making sure the broth covers the vegetables all the way. Place over high heat and bring to a boil. Reduce the heat to maintain a simmer and simmer until very tender, 25 to 30 minutes.

Meanwhile, in a small saucepan, heat the remaining garlic in the oil over low heat and cook until the garlic is tender, about 8 to 10 minutes.

Drain the cooked veggies, transfer to a food processor, and process until smooth, streaming in the garlicky oil and garlic cloves through the hole in the lid as you go until a light, fluffy mash is formed. Season with salt and pepper and serve.

Olive Oil–Poached Carrots and Carrot Sauce

It might seem a bit unusual to poach carrots, but it really concentrates their flavor, imparting earthy, sweet, and slightly creamy qualities. As an added bonus, you end up with a super-flavorful carrot oil—so don't throw out that cooking oil! Save it to make the sauce below to blend into a vinaigrette or to saute veggies in.

Serves 4

1½ pounds small carrots, peeled, with tops on
3 cups extra-virgin olive oil, plus more if needed
1 shallot, thinly sliced
2 cloves garlic, roughly chopped

1 sprig thyme
1 teaspoon whole black peppercorns
1½ teaspoons salt
¼ cup roughly torn fresh mint

To turn carrots into carrot sauce:
1 cup chopped oil-poached carrots (see above)
¼ cup poaching oil (see above)
½ teaspoon salt

¼ cup canned unsweetened coconut milk
¼ cup nutritional yeast
¼ teaspoon mild curry powder

Preheat the oven to 300°F.

Place the carrots in a large baking dish in a single layer. Cover with the oil, making sure the oil goes at least ½ inch over the carrots. Add the shallot, garlic, thyme, and peppercorns and bake for 45 minutes to 1 hour, until the carrots are very tender and the oil reaches 240°F.

Carefully remove the carrots from the oil, saving the oil (see below).

Chop the carrots into bite-sized pieces, add the salt, and toss with the mint before serving, or skip the mint and turn the carrots into carrot sauce.

TO TURN BRAISED CARROTS INTO SAUCE:

In a high-speed blender, combine the carrots, carrot poaching oil, salt, coconut milk, nutritional yeast, and curry powder and blend until smooth and creamy.

Serve as a dipping sauce for vegetables, to saute vegetables in, as the base of a red wine vinaigrette, or as a sauce for seafood, poultry, or lamb.

Seared Salmon Ceviche in Endive Cups

A while back I got hooked on the shrimp ceviche tacos served at a restaurant in LA, and I was inspired to try a healthier twist on the classic. Look for sashimi-grade salmon at a reputable fishmonger or a well-stocked Asian grocery store in your area. If you can't find sashimi-grade salmon, consider lightly cooking your fish (to about 120°F) and then relying on the marinade to "cook" it further. Can't find endive? Use butter or Bibb lettuce leaves instead.

Serves 4

2 heads Belgian endive
1 pound sashimi-grade salmon fillets, skin removed
1 teaspoon salt, divided
½ teaspoon ground cumin
½ teaspoon ground black pepper
2 tablespoons extra-virgin olive oil, divided
1 small red onion, diced

1 medium-sized ripe mango, diced
1 small ripe tomato, diced
1 jalapeño, minced
Juice of 3 limes
Juice of 1 lemon
2 large avocados, diced
¾ cup fresh cilantro leaves

Carefully separate the heads of the endive into individual leaf "cups" and set aside.

Season the salmon with ½ teaspoon of the salt, the cumin, and black pepper.

Heat 1 tablespoon of the oil over medium-high heat. Add the salmon and cook for 1 to 2 minutes, until a golden brown crust forms (it will not be cooked through—this is a raw preparation). Remove the salmon from the heat and let it come to room temperature. Dice the salmon and transfer to a bowl. Add the remaining 1 tablespoon oil, the remaining ½ teaspoon salt, the red onion, mango, tomato, jalapeño, lime juice, and lemon juice. Toss to combine, then let rest in the refrigerator for 10 to 15 minutes, until the salmon starts to lighten and go opaque around the edges. Gently fold in the avocados and cilantro, being careful not to crush them.

Spoon into the endive cups and serve immediately. Because this is a raw preparation, it is important to eat it right away or to keep it chilled until serving (no more than a couple hours after preparing).

CHAPTER 5

MAINS

These mains are hearty and savory and contain ample protein and micronutrients. They'll leave you feeling like a million bucks. Pair with sides from the previous chapter or let them fly solo—either way your mouth and body will be delighted!

Bacalao with Sweet Potatoes, Onions, Peppers, and Leafy Greens

My mother had just a handful of signature dishes that she'd make for me, and this is one of them, which gives it a special place in my heart. This is a variation on a traditional Portuguese bacalao (salted codfish) dish that calls for potatoes and plenty of olive oil. In this variation, I swapped the traditional white potatoes for sweet potatoes, which add a delicious contrast to the salty fish and briny olives. For an added kick, sprinkle with a little red wine vinegar just before serving.

Serves 4

1 pound bacalao (salt cod)*

2 cups white sweet potato, peeled and sliced in 1/4 inch rounds**

3/4 cup extra-virgin olive oil, divided, plus more for the pan

2 medium-sized yellow onions, thinly sliced

4 cloves garlic, thinly sliced

1 cup roughly chopped sweet green bell pepper

1/4 cup roughly chopped jalapeño peppers, seeds and stems removed

4 cups shredded kale leaves

1 pinch salt, or to taste

1/4 teaspoon garlic powder

1/2 teaspoon sweet paprika

1/2 teaspoon smoked paprika

1/2 cup kalamata olives, pitted

1/2 cup roughly chopped fresh parsley

*Bacalao is a salted cod traditional to Spanish, Portuguese, and Brazilian cuisine. It is available in many specialty and gourmet shops and also on Amazon. If you can't find bacalao, try this dish with smoked trout—skip the soaking in the first step. The flavor won't be the same, but it still will be incredibly delicious.

**If you can't find white sweet potatoes, it's fine to use orange sweet potatoes.

Soak the bacalao in a large bowl of water for 24 to 36 hours, changing the water occasionally (at least six changes of water). Drain.

Break the bacalao into flakes, removing any large bones. Set aside.

Put the sweet potatoes in a large pot and cover with water. Bring to a boil and cook over medium heat for about 20 minutes, until tender. Drain and rinse under cold water.

Preheat the oven to 350°F. Lightly brush a large baking dish with oil and set aside.

Meanwhile, in a large saute pan, heat 1/4 cup of the oil over medium heat. Add the onions and cook for 3 to 4 minutes, until tender. Add the garlic and peppers and continue to cook, stirring occasionally, until the peppers are tender, 2 to 3 minutes. Add the kale and salt and cook until the kale is just wilted, 1 to 2 minutes.

In a small bowl, stir together remaining ½ cup oil with the garlic powder, sweet paprika, and smoked paprika. Pour the seasoned oil over the flaked fish mixture and toss to coat.

Spread half of the potatoes over the baking dish. Add half of the vegetable mixture and half of the bacalao. Repeat the layering and top the final layer of bacalao with the olives. Bake for 20 to 25 minutes, until lightly browned and warmed through. Top with the parsley and serve.

Simmered Ginger Salmon with Daikon Radish and Greens

Inspired by traditional nimono, or Japanese simmered dishes, this features an interesting combination of sweetness from the radish and miso, richness from the salmon, and a bit of a bite from the mustard greens. It's great over sauteed cauliflower rice.

Serves 4

1 small daikon radish root (about 1 pound), cut into bite-sized chunks

4 cups water

4 tablespoons white or red miso paste

1 (3-inch) chunk ginger, peeled and roughly chopped

1 pound skinless salmon, cut into large chunks

2 cups roughly chopped napa cabbage

2 cups roughly chopped mustard greens

Hot sauce (chili garlic sauce is great for this)

Coconut aminos or tamari sauce

Put the daikon in a pot with a tight-fitting lid. Add the water and miso paste and cook over medium heat, stirring frequently, until the daikon is tender and the miso is dissolved, about 20 minutes. Add the ginger and cook for an additional 5 minutes, or until fragrant.

Add the salmon, cabbage, and mustard greens to the pot and bring to a simmer. Reduce the heat to low, cover, and simmer for 10 to 12 minutes, until the greens are wilted and the salmon begins to flake. Taste and adjust the heat with hot sauce or the salt/umami with coconut aminos. Serve as is or over sauteed cauliflower rice.

Olive and Olive Oil–Braised Chicken and Greens with Orange and Fennel

Fennel, apples, oranges, and olives are the perfect flavor combination, whether raw or cooked, and the hint of ginger adds a bit of unexpected heat. I love braising chicken this way, not just because you get a great one-pot meal, but because the flavors of all the vegetables meld beautifully, making a vegetable "sauce" for your chicken.

Serves 4

4 tablespoons extra-virgin olive oil, divided

4 skin-on chicken thighs

1 teaspoon salt, divided

2 whole leeks, cleaned and thinly sliced

1 bulb fennel, cored and thinly sliced (discard fronds)

1 Granny Smith apple, peeled, cored, and cubed

1 tablespoon minced fresh ginger

1 tablespoon minced garlic (2 to 3 cloves)

1 teaspoon ground black pepper

1 teaspoon cumin seeds

1 teaspoon paprika

3 cups shredded mustard greens or collard green leaves

3 large oranges, cut into segments

1 cup pitted green olives in brine

Preheat the oven to 375°F.

In a large oven-safe saucepan, heat 2 tablespoons of the oil over medium-high heat. Season the chicken thighs with a pinch of the salt and transfer to the pan skin-side down. Cook for 5 to 7 minutes, until the skin is very golden brown and crisp, then remove from the heat to a plate and set aside (the chicken should not be fully cooked at this point).

Add the leeks, fennel, and apple to the pan and season with the remaining salt. Increase the heat to high and cook for 6 to 7 minutes, until leeks and fennel are very tender. Add the ginger, garlic, black pepper, cumin seeds, and paprika and cook for an additional 2 to 3 minutes, until very fragrant. Add the greens and cook until they are thoroughly wilted, about 4 to 5 minutes. Remove from the heat and fold in the oranges, olives, and olive brine.

Return the chicken to the pan skin-side up. Transfer to the oven and cook for 30 to 35 minutes, until the chicken is falling-off-the-bone tender. Remove from the oven and serve.

Bone Broth Beef Stew with Purple Sweet Potatoes

A good stew warms the soul, so what could be better than that? A bone broth stew that combines the silky mouthfeel of collagenous broth with butter-soft beef cubes. I've added purple sweet potatoes to balance out the savory beef and lend a hit of heart-healthy potassium and brain-boosting anthocyanins. If you can't find purple sweet potatoes, feel free to use the usual orange; it'll be just as delicious!

Serves 4

4 teaspoons salt

1½ pounds boneless beef chuck, cut into 1-inch cubes

1 tablespoon extra-virgin olive oil

3 cups beef bone broth

4 medium carrots, chopped

1 medium yellow onion, diced

1 pound purple sweet potatoes, cut into bite-sized pieces

3 cloves garlic, minced

2 tablespoons tamari sauce (coconut aminos can also be used)

1 tablespoon balsamic vinegar

2 tablespoons tomato paste

1 teaspoon dried oregano

2 teaspoons dried thyme

2 teaspoons paprika

½ teaspoon ground black pepper

Season the beef all over with 3 teaspoons of the salt.

Heat the oil in a large soup pot and sear the beef until evenly browned on all sides, about 10 minutes.

Add the bone broth, carrots, onion, sweet potatoes, garlic, tamari, vinegar, tomato paste, oregano, thyme, paprika, black pepper, and the remaining 1 teaspoon salt and stir well. Bring to a boil over high heat, then reduce the heat to maintain a low simmer and cook covered for 1½ hours. Remove from the heat, let cool slightly, and serve.

Bangin' Liver

This is a recipe from my amazing friend Mary Shenouda, aka @PaleoChef on Instagram. She's a private performance chef to elite athletes and entertainers and believes liver is a superfood. Of course, I concur! I'd never tasted chicken liver before trying Mary's dish years ago, but it made me and makes anyone I cook it for an instant convert. It is easy to prepare, utterly delicious, and I promise you'll feel charged up after eating it from the many nutrients it contains, including choline, vitamin B_{12}, folate, and vitamins A and K_2.

Serves 2 to 3

1 pound chicken livers, cleaned and roughly chopped
¾ teaspoon salt
⅓ cup ghee, plus more for serving
6 cloves garlic, minced
1 large green bell pepper, chopped
1 jalapeño, seeded and chopped

1 tablespoon ground cumin
½ teaspoon ground cinnamon
¼ teaspoon ground ginger
¼ teaspoon ground cloves
¼ teaspoon ground cardamom
Juice of 1 lime, plus more for serving
2 tablespoons fresh cilantro, chopped

Sprinkle the livers with the salt, give them a toss, and set aside for 2 to 3 minutes.

Heat the ghee in a large skillet over medium-high heat. Add the liver and sear until browned on both sides, about 4 minutes per side. Add the garlic, bell pepper, and jalapeño and cook until the vegetables are starting to soften, about 5 minutes. Add the cumin, cinnamon, ginger, cloves, and cardamom, reduce the heat to medium-low, cover, and cook for another 5 to 8 minutes, until a thermometer inserted into the thickest part reaches 160 degrees.

Add the lime juice and scrape up any browned bits from the bottom of the pan, mixing them in. Remove from the heat. Serve with an additional hit of melted ghee, a touch of lime juice, and cilantro to garnish.

Ghee-Seared Beef Liver

I consider beef liver to be a superfood, loaded with vitamin A, B vitamins, selenium, vitamin K_2, and copper. Unfortunately, for most who've attempted to make it, cooking beef liver is not very intuitive, and the results can quickly ruin a first impression. Well-cooked liver should have a nice, crispy crust on the outside but remain rare on the inside, lending it an almost creamy texture. Using a generous hit of salt, an acid (like lemon), and ghee really softens the strong flavor of liver, making it delicious. I try to eat a few ounces of liver every week, and this is my go-to recipe!

Serves 1

4 ounces beef liver

1½ teaspoons coarse or fine salt

1 tablespoon ghee

¼ lemon

Sprinkle the salt on both sides of the liver. A generous, evenly distributed pinch on each side should do.

Heat the ghee in a pan over high heat. Once the pan is very hot, place the liver in the pan and cook for about 1 minute, until when you peek on the bottom you see a nice sear developing. Flip the liver and sear the other side for about another minute. If the liver is browned but has not yet developed a slight "crust," flip it again and cook until it has.

Remove from the heat and let sit for 2 minutes. Slice the liver, squeeze the lemon over it, and serve.

Insanely Crispy Gluten-Free Buffalo Chicken Wings

Most chicken wings are unhealthy—feedlot animal parts fried in unhealthy oils and breaded with refined flour (yuck!). These, however, are baked, grain-free, and full of nutrients. Chicken skin is full of collagen, as are the cartilage-rich joints of the chicken wing. Plus, dark meat chicken is a great source of heart-healthy vitamin K_2. When shopping for your hot sauce, make sure that it contains only red pepper, vinegar, salt, and garlic.

Serves 2 to 3

Softened or melted coconut oil

1 pound chicken wings

Garlic salt (I like Redmond Real Salt Organic Garlic Salt)

½ cup hot sauce (I like Frank's RedHot Original Cayenne Pepper Sauce)

2 tablespoons butter

Cayenne pepper (optional)

Preheat the oven to 250°F and grease a sheet pan with coconut oil.

Place the wings on the prepared sheet and sprinkle with garlic salt. Give them a nice even seasoning (one side is fine).

Bake the wings for 45 minutes. Why such a low temperature? It helps dry out the wings and melts away extra fat and connective tissue. Very important! (Note: the wings are not done after this step—do not eat yet!)

Turn the heat up to 425°F and bake for another 45 minutes, or until the wings have a nice golden brown color and have shrunk considerably. Remove from the oven and let sit at room temperature for 5 minutes.

While the wings are resting, combine the hot sauce, butter, and extra cayenne, if using, in a small saucepan over very low heat just to warm the hot sauce and melt the butter.

Whisk the wing sauce, then transfer to a large bowl or pot. Throw in the wings, toss well to coat them with the sauce, and serve.

Olive Oil Lamb Chops with Olives and Artichoke Hearts

Lamb and artichokes are a classic combination for a great reason—the gaminess of the lamb pairs beautifully with the briny artichoke hearts. The tomatoes add a nice burst of sweetness to this super flavorful dish, and the chile flakes add just a bit of heat without making this hearty dinner too spicy. Okey dokey, artichokey!

Serves 4

4 lamb shoulder chops (½ inch thick)
1½ teaspoons salt, divided
1 teaspoon ground black pepper
4 tablespoons extra-virgin olive oil
1 red onion, thinly sliced
4 cloves garlic, thinly sliced
2 tablespoons minced fresh rosemary
½ teaspoon red chile flakes
3 tablespoons tomato paste

1 (14-ounce) can artichoke heart quarters, drained and thoroughly dried
1 cup cherry tomatoes
Juice of 1 lemon
1 cup pitted niçoise olives
About 1 cup chicken bone broth, plus more as needed
¼ cup minced fresh parsley

Season the lamb chops front and back with ½ teaspoon of the salt and the black pepper.

Heat 2 tablespoons of the oil in a pan over medium-high heat, add the lamb chops, and sear for 2 to 3 minutes per side for medium-rare. Remove from the pan to a plate.

Add the remaining 2 tablespoons oil to the pan along with the onion, garlic, rosemary, and chile flakes and cook for 3 to 5 minutes, until the onion is tender and the mixture is very fragrant.

Add the tomato paste and artichoke hearts and cook, stirring frequently, until the artichoke hearts are browned and the tomato paste is slightly toasted, about 8 minutes. Add the cherry tomatoes, lemon juice, olives, and ⅓ cup of the chicken broth and stir to combine. Bring to a simmer and simmer for 5 to 10 minutes, adding broth as needed until a flavorful sauce has formed and the cherry tomatoes are bursting.

Season with the remaining 1 teaspoon salt and sprinkle with the parsley. Serve each lamb chop with a generous helping of the vegetable mixture.

Beef Picadillo

I lived in Miami, a hub of epic Cuban food, for four years and couldn't get enough of a good picadillo. Here's a healthy variation on this traditional dish that I make often. It goes great over sauteed cauliflower rice or, if you really want to keep the vampires away, the garlicky zucchini noodles on page 130.

Serves 2 to 3

1 tablespoon extra-virgin olive oil
1 large yellow onion, finely chopped
4 cloves garlic, smashed and peeled
1 pound ground beef
1 teaspoon salt

1½ teaspoons ground black pepper
¼ teaspoon red chile flakes (optional)
½ cup no-added-sugar tomato sauce
½ cup pitted green olives, sliced (olives stuffed with pimientos are fine)

Heat the oil in a large skillet over medium heat. Add the onion and cook for 4 to 5 minutes, until softened. Throw in the smashed garlic and allow to aromatize for 1 minute. Add the ground beef, throw on the salt, pepper, and red chile flakes, if using, and cook, stirring often to break it up, until browned, about 10 minutes. Add the tomato sauce and olives, bring to a simmer, then reduce the heat to very low and simmer for 10 minutes. Serve.

Garlicky Olive Oil Shrimp

My mom wasn't kosher but did have some food rules derived from her faith. For one, she never knowingly allowed shrimp in the house. Unfortunately, that never stopped me or my brothers from sneaking it in (sorry, Mom). The delicious taste and texture provided by crustaceans like shrimp is simply undeniable, surpassed only by the nutrition they offer. This recipe is sort of a cross between New Orleans–style barbecue shrimp and a classic Italian American shrimp scampi, in all the best ways. It's strongly spiced, nice and garlicky, and perfect over sauteed cauliflower rice or, if you really like garlic, the zucchini noodles on page 130.

Serves 4

¼ cup extra-virgin olive oil

1 head garlic, cloves thinly sliced

1½ pounds large shrimp, peeled and deveined (tail-on is OK) *

1 teaspoon salt

1 tablespoon Old Bay seasoning

½ teaspoon paprika

Zest and juice of 1 lemon

¼ cup chopped fresh parsley

¼ cup nutritional yeast

*If using frozen, make sure to thaw completely before cooking.

Place the oil and garlic in a large saute pan over medium heat and cook, stirring occasionally, until the garlic is very fragrant and light golden brown, about 2 minutes. Remove the garlic from the oil using a slotted spoon.

Turn the heat up to high, add the shrimp to the oil, and cook for 1 to 2 minutes, until shrimp is just beginning to look opaque. Reduce the heat to medium and add the salt, Old Bay, paprika, and lemon zest. Cook for an additional 1 to 2 minutes, until the shrimp are *just* cooked (you don't want them to get tough). Add the lemon juice, parsley, and nutritional yeast, then toss with the crispy garlic and serve.

Perfect Ribs

Ever since I was a teenager, I've adored a good rack of ribs. Smoky, fatty, sweet, and salty—is there anything more decadent? Unfortunately, it's hard to find ribs that aren't also covered in a sugary sauce. Like any collagen-rich body part, the key to making ribs that melt in your mouth is to cook them slow, which ensures that the collagen will melt down to gelatin. Gelatin is what gives ribs (and properly cooked chicken drumsticks, lamb shoulder, or any other joint) that silky, moist texture. For these ribs, I've created a rub combining sweet and savory—but without the sugar—that's so good you won't even need sauce. You can also use this rub on pork tenderloin, roasted chicken, salmon, or any other protein your heart desires.

EPIC RIB RUB

Makes about ¼ cup of rub

1½ teaspoons sugar-free "brown" sweetener*
1½ teaspoons garlic powder
1½ teaspoons onion powder
1½ teaspoons chili powder
1½ teaspoons smoked paprika
½ teaspoon ground black pepper
¼ teaspoon cayenne pepper
1 teaspoon salt

* For this recipe, I use Lakanto brand Golden monk fruit sweetener. If you can't find it, look for a sugar-free brown sugar equivalent. Otherwise, regular "white" monk fruit sweetener is fine. This assumes a one-to-one sweetness ratio of sugar to sugar alternative. Make sure that whatever sweetener you use, the sweetness equivalent is to 1½ teaspoons sugar.

In a small bowl, combine the sweetener, garlic powder, onion powder, chili powder, smoked paprika, black pepper, cayenne pepper, and salt. Mix well.

EPIC RIBS

Serves 2 to 3

Epic Rib Rub (page 202)
1 rack baby back ribs

Preheat the oven to 275°F and line a sheet pan with parchment paper.

Rub the ribs down with the rub on the outwardly curving side, leaving ample sitting on top.

Place on the prepared sheet pan outwardly curving–side up and bake uncovered for 3½ hours, until meat is tender, and rib bones come away from meat when pulled firmly.

Note: Some people like to remove the membrane on the underside of the ribs, but I don't mind it, and I find it adds a fun texture. But if you'd like truly fall-off-the-bone ribs, before seasoning, flip the ribs over, identify the membrane, slide a dinner knife under it to release, and then peel off.

Coconut Braised Chicken Thighs with Mustard Greens and Cauliflower

I love chicken thighs; not only are they one of the more flavorful parts of the chicken, they're economical and provide nutritional benefits not found in the breast. Namely, they're a great source of collagen as well as vitamin K_2 (chicken dark meat contains six times the vitamin K_2 of breast meat). The sweetness of the coconut milk plays perfectly with the sharp bite from the greens, rounding out this hearty braise. Cauliflower adds body to the dish, making it an ideal one-pan meal for a busy evening.

Serves 4

2 tablespoons avocado oil

4 skin-on chicken thighs

1 teaspoon salt, divided

4 cloves garlic, roughly chopped

2 leeks, cleaned and thinly sliced

1 medium-sized head cauliflower, cut into small florets

Zest and juice of 1 lemon

1 teaspoon ground black pepper

½ teaspoon ground turmeric

½ teaspoon paprika

½ teaspoon ground cumin

1 bunch mustard greens, stems removed, leaves very thinly sliced

1 (15-ounce) can full-fat unsweetened coconut milk

2 tablespoons unsweetened coconut flakes (optional)

Preheat the oven to 375°F.

In a large oven-safe saucepan, heat the oil over medium-high heat. Season the chicken thighs with a pinch of the salt and place in the hot pan skin-side down. Cook 5 for 7 minutes, until the skin is very golden brown and crispy, then remove from the pan and set aside on a plate (the chicken should *not* be fully cooked at this point).

Add the garlic, leeks, and cauliflower to the pan, then add the remaining salt, the lemon zest, black pepper, turmeric, paprika, and cumin. Increase the heat to high and cook for 5 to 7 minutes, until the cauliflower begins to crisp around the edges and the leeks are tender. Add the mustard greens and cook until just wilted, 2 to 3 minutes. Add the coconut milk and lemon juice.

Place the chicken skin-side up in the veggie mixture, transfer to the oven, and bake for 30 to 35 minutes, until the chicken is falling-off-the-bone tender. Remove from the oven, sprinkle with coconut flakes, and serve.

Calo Verde (Portuguese-Style Greens and Sausage Stew)

I had the pleasure of visiting Portugal a few years ago and fell in love with the cuisine. This garlicky soup is a twist on a classic. Not only is it delicious, it's a great way to use up hearty greens in your fridge. The combination of greens and sweet potatoes results in a creamy soup studded with crispy, savory sausage. If you love this, definitely try the bacalao on page 184.

Serves 4 to 6

¼ cup extra-virgin olive oil

8 ounces linguiça, chorizo, or other hard, garlicky sausage, cut into thin coins

1 medium yellow onion, minced

4 cloves garlic, minced

1 teaspoon salt, or to taste

1½ teaspoons ground black pepper

1 pound white sweet potatoes, peeled and diced *

4 cups water

4 cups chicken bone broth or vegetable broth

1 pound collard greens, stems removed, leaves cut into bite-sized pieces

*You can also use orange sweet potatoes, but white sweet potatoes will keep the finished product a more vibrant green.

In a large heavy-bottomed soup pot, heat the oil over medium-high heat. Add the sausage and cook until crisp, about 4 to 6 minutes, then remove with a slotted spoon to a plate and set aside for later.

Add the onion to the pot and cook, stirring frequently, until tender, about 5 minutes. Add the garlic, salt, black pepper, and sweet potatoes and continue to cook for 3 to 5 minutes more, until the garlic is fragrant. Add the water and broth, bring to a simmer, then reduce the heat to maintain a simmer and cook until the sweet potatoes are tender, 15 to 20 minutes.

Using a hand blender, blend the soup until smooth. Add the collard greens, increase the heat, bring to a boil, and cook until the collard greens are wilted, about 10 minutes. Add the sausage, stir, adjust the seasonings as needed, and serve.

Braised Chicken with Artichokes, Fennel, and Kale

Another dish featuring a few of my favorite classic Mediterranean flavors, this hearty one-pan meal is perfect for a busy worknight dinner. If you're not a fan of fennel, consider substituting celery or thinly sliced leeks—but give this a try first. It's changed a lot of fennel haters' minds, thanks to the addition of tangy artichokes and perfectly crisp-skin chicken.

Serves 4

2 tablespoons avocado oil

4 skin-on chicken thighs

1 teaspoon salt, divided

2 medium red onions, thinly sliced

1 bulb fennel, cored and thinly sliced

1 (14-ounce) can quartered artichoke
   hearts, drained and patted dry

Zest of 1 orange

Zest and juice of 1 lemon

1 teaspoon ground black pepper

1 teaspoon anise seeds

2 tablespoons nutritional yeast

1 tablespoon Dijon mustard

1 bunch kale, stems removed, leaves very
   thinly sliced

2 cups chicken bone broth

Preheat the oven to 375°F.

Heat the oil in a large oven-safe saucepan over medium-high heat.

Season the chicken thighs with a pinch of the salt and place in the hot pan skin-side down. Cook for 5 to 7 minutes, until the skin is very golden brown and crisp, then remove from the pan to a plate and set aside (the chicken should *not* be fully cooked at this point).

Add the onions, fennel, and artichoke hearts to the pan and season with the remaining salt. Increase the heat to high and cook for 4 to 5 minutes, until onions and fennel are tender, and artichokes are beginning to brown. Add the orange and lemon zests, the black pepper, and anise seeds and cook for an additional 2 to 3 minutes, until very fragrant. Add the nutritional yeast, mustard, and kale and cook until the mixture is well combined and the kale is wilted, 3 to 5 minutes.

Return the chicken to the veggie mixture skin-side up and carefully pour the broth into the pan, making sure not to cover the skin (you want the skin to crisp back up). Transfer to the oven and bake for 30 to 35 minutes, until the chicken is falling-off-the-bone tender. Remove from the oven, squeeze lemon juice over the vegetables, and serve.

The Perfect Steak

Knowing how to make a steak is essential. It's also so easy to do. This method yields a restaurant-quality steak with the most basic setup: a cast-iron pan. In terms of cut, I love cooking a delicious marbled rib eye or a lean and tender tenderloin (aka filet mignon). No matter your choice, don't cut any fat off the steak; it will come in handy once you start cooking.

Serves 1 to 2

Coarse (kosher) salt
1 steak
1 teaspoon ghee or avocado oil*
1 pinch flake salt

* If you're cooking a fatty steak, you do not need to add any oil to the pan. A leaner cut (tenderloin, for example) would require some oil in the pan.

Set your steak on the countertop to allow it to come to room temperature, usually about 30 minutes. Just before cooking, blot steak on all sides with a paper towel to dry the surface. Any water on the steak's exterior will prevent a great crust from forming.

Set a cast-iron pan over high heat and allow it to get very hot. If you're using oil, put it in the pan before heating. If your steak has a fat cap, put that side down on the pan first (holding the steak upright with tongs) and let it cook for 1 minute. This will render out some of the beef's own fat to cook on. Spread the fat around the pan.

Place the steak flat on the pan and sprinkle evenly with coarse salt, saving an equal amount for the other side. Allow it to develop a nice crust, typically 2 to 3 minutes. Flip the steak. Continue to flip until you get a nice, even sear.

Gently poke top of the steak with your finger (careful!—it'll be hot). For medium-rare, the steak should feel like the skin at the base of your thumb when touching your thumb to your middle finger. If it's softer, cook it a little more. (You can also use an instant-read thermometer; aim for an internal temperature of 125°F for medium-rare and then pull from direct heat).

Remove from the pan, place on a cutting board, and allow the steak to rest for 5 to 7 minutes. Slice and finish with flake salt.

Note: For a simple Tuscan variation, before finishing with the flake salt, drizzle with 1 tablespoon extra-virgin olive oil and ½ teaspoon balsamic vinegar.

"Pattyless" Jamaican Beef Patty

When I was a kid in New York, one of my absolute favorite after-school snacks was the Jamaican beef patties I'd buy from the local pizza shops. Delicious as they were, they probably were loaded with trans fats, refined grains, and industrial oils. Here, I've re-created the seasoning of the beef, and the nutritional yeast lends the flavor of the warm, doughy crust. This is one nourishing dish and goes great with my "Cheesy" Kale Salad (page 159).

Serves 2 to 3

1 teaspoon ghee
1/2 medium yellow onion, chopped
5 cloves garlic, smashed and peeled
1 pound ground beef
1 teaspoon salt
1 tablespoon ground cumin
1 1/2 teaspoons ground turmeric

1/2 teaspoon ground coriander
1/2 teaspoon ground allspice
1/2 teaspoon ground cardamom
1/4 teaspoon ground black pepper
1/4 cup nutritional yeast (optional but recommended)

Heat the ghee in medium skillet over medium heat. Add the onion and cook for 4 to 5 minutes, until softened. Throw in the smashed garlic and allow to aromatize for 1 minute.

Add the ground beef, throw on the salt and all the spices, and cook, stirring often to break it up, until the beef is browned, about 10 minutes. Sprinkle on the nutritional yeast and serve.

Simple Burger Patty

To make the best burger you've ever had, less is more. Instead of mixing salt, spices, and, God forbid, raw veggies into your ground beef (you really expect raw onions to cook and caramelize in the middle of a patty?), you're going to sprinkle salt on the surface of your patty just before grilling. Here's why: salt changes the texture of the protein in meat, and mixing it in before grilling will give your burger a more sausage- or meatball-like consistency. This recipe really lets the meat be the star and is one of my staple recipes, because a burger patty can be thrown on top of any salad or greens dish to complete the meal. Plus, 100% grass-fed ground beef is so economical. When cooking fatty meat like ground beef, you won't need to add oil to the pan provided it's very hot when you put the meat in.

Serves 2 to 4

1 pound ground beef (85 percent lean or higher)

1 tablespoon coarse salt

Set a large skillet over medium-high heat.

Form the beef into 2 to 4 burger patties (the size is up to you), making the patties as flat as possible. Add a small indent in the center with your thumb (this helps them hold their shape).

Sprinkle the salt on each side of the patties, keeping your fingers about 8 inches above the patties to ensure even distribution and no salty "hot" spots.

Place the patties in the hot pan and cook for 4 to 5 minutes, then flip and cook for an additional 3 to 4 minutes, until outside of burger is crisp and brown all over. Serve on a bed of greens, or just eat as is.

Lemongrass Salmon Banh Mi Bowl

Vietnamese food is among my favorite cuisines. Inspired by the classic sandwich, this bowl features a delicious blend of fresh, tangy, sweet, and spicy flavors to keep your taste buds excited for every bite. As an alternative, it's worth trying it with grass-fed beef or pasture-raised pork instead of the salmon—and you can pickle any vegetable you'd like along with the carrots, shallot, and jalapeño—radishes, broccoli stems, even garlic work beautifully.

Serves 2

¾ cup rice vinegar, divided
2 tablespoons monk fruit sweetener, divided
2 tablespoons coconut aminos or tamari
2 medium carrots, cut into matchsticks
1 large shallot, thinly sliced
1 jalapeño, thinly sliced
8 ounces skin-on salmon fillet
1 stalk lemongrass, crushed and finely chopped

1 tablespoon grated fresh ginger
2 tablespoons avocado mayonnaise
1 tablespoon hot sauce
Juice of 1 lime
4 cups mixed greens
1 avocado, thinly sliced
¼ cup minced fresh parsley
¼ cup minced fresh cilantro

First, make the pickled veggies: Heat ½ cup of the vinegar, 1 tablespoon of the monk fruit, and the coconut aminos in a small saucepan over medium heat. Add the carrots, shallot, and jalapeño, stir, and remove from the heat. Let sit for 30 minutes.

Meanwhile, preheat the oven to 375°F. Place the salmon in an oven-safe dish.

In a small bowl, whisk together remaining ¼ cup vinegar, remaining 1 tablespoon monk fruit, the lemongrass, and ginger. Pour the mixture over the salmon and bake for 15 to 18 minutes, until the salmon is tender and flakes with a fork. Let the salmon cool to room temperature.

Right before serving, make a quick sauce by whisking together the avocado mayonnaise, hot sauce, and lime juice in a small bowl.

Serve the salmon over the greens garnished with the pickled veggies, avocado, and fresh herbs. Drizzle the sauce over the top or serve it on the side.

Spice-Rubbed Salmon with Almond Basil Pesto

There are a million variations on pesto out there, including a few in this book. This particular recipe, utilizing almonds, goes especially well with salmon. The arugula adds a nice peppery bite, and the tang of the lemon zest pairs beautifully with the salmon. Any leftover pesto can be tossed with sweet potato noodles, stirred into your favorite dressing, or used to top steamed broccoli.

Serves 2 to 4

For the pesto:

2 cups fresh basil leaves

1 cup arugula leaves

2 cloves garlic, peeled

½ cup toasted almonds

Zest of 1 lemon

½ teaspoon salt, or to taste

¾ cup nutritional yeast

½ cup extra-virgin olive oil

For the salmon:

1 pound skin-on salmon fillets, cut into 4 equal pieces

2 tablespoons avocado oil, divided

1 teaspoon salt

½ teaspoon ground black pepper

½ teaspoon ground cumin

½ teaspoon paprika

¼ teaspoon ground cloves

½ batch pesto (see above)

TO MAKE THE PESTO:

In a food processor, pulse together the basil, arugula, garlic, almonds, lemon zest, salt, and nutritional yeast until well combined. With the motor running, gradually stream the oil in through the hole on the top until all the oil is incorporated. Taste and add more salt as needed.

You can store the pesto in a covered container in the refrigerator for up to 1 week or in the freezer for up to 6 months.

TO MAKE THE SALMON:

Preheat the oven to 400°F.

Spread 1 tablespoon of the oil over the salmon.

In a small bowl, stir together the salt, black pepper, cumin, paprika, and cloves. Coat the meat (non-skin) side of the salmon with the spice mixture.

Heat the remaining 1 tablespoon oil an oven-safe skillet over medium-high heat. Add the salmon to the pan skin-side down and cook until the skin is crispy and easily releases from the pan, about 5 minutes.

Transfer to the oven and bake for 15 minutes. Remove from the oven and spoon half of the pesto over the salmon. Return to the oven and bake for 5 minutes, or until an instant-read thermometer inserted into the thickest part of the salmon reads 125°F. Serve with the remaining pesto.

Miso Sesame Roasted Salmon with Broccoli, Garlic, and Cabbage

I'm a big fan of sheet pan suppers for their simplicity—they're great weeknight meals, they don't lead to a huge cleanup, and when they're made using a super flavorful, savory marinade like this one, they're downright addictive. This recipe was inspired by the shrimp broccoli at a local Chinese restaurant, and is great with cauliflower rice if you find you need a little something extra.

Serves 2 to 4

1/3 cup miso paste (white or red works best)

2 tablespoons coconut aminos or tamari sauce

1 tablespoon rice vinegar

2 tablespoons monk fruit sweetener

3 tablespoons grated fresh ginger, divided

2 skin-on salmon fillets (about 8 ounces each)

2 cups bite-sized broccoli florets (1/2 large head)

1/2 small head red cabbage, cored and chopped into bite-sized chunks

1 head garlic, cloves roughly chopped

2 tablespoons sesame oil

1/2 cup thinly sliced scallions

1 tablespoon sesame seeds

Preheat the oven to 400°F. Line a rimmed sheet pan with parchment (optional, but makes for easier cleanup) and set aside.

In a small bowl, whisk together the miso paste, coconut aminos, vinegar, monk fruit, and 2 tablespoons of the ginger.

Place the salmon skin-side down on sheet pan and brush with half of the miso mixture. Reserve the remaining sauce.

In a large bowl, toss together the broccoli, cabbage, and garlic. Add the oil and remaining 1 tablespoon ginger and toss to coat the veggies. Scatter the vegetable mixture around the salmon on the sheet pan.

Bake for 20 minutes, or until the salmon is cooked through (it should flake easily with a fork) and the veggies are tender. Remove from the oven and toss the veggies with the remaining miso mixture. Sprinkle with the scallions and sesame seeds and serve.

Braised Lamb with Broccoli Salsa in Butter Lettuce "Tortillas"

These tacos are inspired by traditional barbacoa—rich, spiced braised lamb. I've made the less-than-traditional choice to pair these delicious tacos with a broccoli-based salsa to add much-wanted crunch, and creamy coconut yogurt for tang. If broccoli rice isn't available in your grocery store, simply pulse the stems of broccoli in a food processor until broken into rice-sized pieces.

Serves 4

For the lamb:

Juice of ½ tangerine

Zest and juice of 1 lime

1 teaspoon ground cumin

½ teaspoon ground cloves

4 cloves garlic, crushed

1 teaspoon salt

2 canned chipotles, chopped (from a can of chipotles in adobo)

1 pound lamb stew meat, cut into 1-inch cubes

3 tablespoons sesame oil

1 large red onion, sliced

3 cups chicken bone broth or vegetable broth

1 sprig rosemary

1 stick cinnamon

1 sprig oregano

Canned unsweetened coconut milk (optional)

For the broccoli salsa:

2 cups broccoli rice

1 large red bell pepper, diced

1 jalapeño, diced

1 Roma tomato, diced

2 large red onions, diced

¼ cup minced fresh cilantro

1 clove garlic, crushed

Juice of 1 lime

Juice of 1 lemon

¼ cup extra-virgin olive oil

½ teaspoon salt

For serving:

1 head butter lettuce, leaves separated

1 cup plain unsweetened coconut yogurt

2 tablespoons nutritional yeast

Lime wedges

TO MAKE THE LAMB:

In a large bowl, combine the tangerine juice, lime juice, lime zest, cumin, cloves, garlic, salt, and chipotles. Add the lamb, stir to coat in the spice mixture, cover, and marinate in the refrigerator for at least 3 hours or as long as overnight.

In a large pot, heat the oil over medium heat. Add the onion and cook until tender, 2 to 3 minutes, then increase heat to medium high, add the lamb, and sear on all sides, ½ minute per side. Add the broth, rosemary, cinnamon, and oregano and bring to a boil. Reduce the heat to low, cover, and simmer for 1 hour.

Remove the lid from lamb and continue to cook until the liquid thickens and the lamb is easy to shred with a fork, about 40 minutes. Shred the lamb, taste, and adjust the salt if needed; if it's too spicy, thin it with a little coconut milk.

TO MAKE THE BROCCOLI SALSA:

While the lamb is cooking, combine all the salsa ingredients in a large bowl and let rest at room temperature for 30 minutes. Refrigerate until needed.

TO SERVE:

Assemble your tacos using lettuce leaves as the base. Fill each leaf with shredded lamb, then top with salsa, a drizzle of coconut yogurt, and a sprinkle of nutritional yeast. Serve with lime wedges.

Salmon, Bok Choy, and Herbs in Coconut Broth

Thai food provides a cornucopia of exciting flavors, balancing sweet, salty, spicy, and acidic. I'm especially enamored with the Thai coconut-based soup tom kha kai, which I must try in any new Thai restaurant I visit. Here's a Thai-inspired poached salmon in a broth that's so flavorful you'll want to drink it on its own. It's delicious served as is, over cauliflower rice, or over mashed sweet potatoes for a heartier dish.

Serves 4

4 (6-ounce) skinless salmon fillets
1 teaspoon salt
2 tablespoons sesame oil
1 large shallot, thinly sliced
5 cloves garlic, thinly sliced
1 (3-inch) piece ginger, peeled and cut into matchsticks
1 jalapeño, finely diced (optional)

2 (15-ounce) cans full-fat unsweetened coconut milk
1 teaspoon monk fruit sweetener
4 cups sliced bok choy
1 bunch green onions, finely chopped
¼ cup chopped fresh cilantro
1 cup lightly torn fresh Thai basil leaves
Coconut aminos or tamari sauce

Season the salmon with the salt and set aside.

Heat the oil in a large pot over medium-high heat. Add the shallot, garlic, ginger, and jalapeño and cook, stirring frequently, until the shallot is tender and the mixture is very fragrant, 3 to 5 minutes. Add the coconut milk and monk fruit, bring to a simmer, then reduce the heat to maintain a simmer for 1 to 2 minutes, until the monk fruit is dissolved. Add the salmon, cover, and cook for 10 to 12 minutes, until the fish is tender. Transfer the fish to a bowl and cover to keep warm.

Add the bok choy and green onions to the pot and cook uncovered until the greens are wilted, about 10 to 15 minutes. Fold in the cilantro and Thai basil and spoon the broth and greens over the fish. Serve with coconut aminos to add to individual bowls.

Japanese-Style Beef, Sweet Potato, Miso, and Mushroom Stew

This rich, savory stew takes inspiration from Japanese nimono, or simmered dishes. I highly recommend using dashi if you can find it—it's available in most well-stocked grocery stores, near the soy sauce. If you can find Japanese sweet potatoes, definitely use them—they're a bit sweeter and creamier than the sweet potatoes most of us grew up with, and the sweetness is really lovely with the beef. That said, it's delicious with the sweet potatoes typically found in the grocery store, too.

Serves 3 to 4

2 tablespoons sesame oil

1½ pounds beef chuck, cut into bite-sized cubes

2 tablespoons red or white miso paste

2 cups sliced mushrooms (shiitake or porcini are best)

2 cups dashi (Japanese fish stock) or chicken bone broth

1 large ginger root, about 6 inches, sliced into rounds (about ½ cup rounds)

¼ cup coconut aminos or tamari

3 tablespoons monk fruit sweetener

2 cups diced sweet potatoes (cut the sweetener in half if using Japanese sweet potatoes)

Juice of 1 lemon

¼ cup sliced green onion

Heat the oil in a large heavy-bottomed soup pot over medium-high heat. Add the beef and sear on all sides, about 10 minutes. Add the miso and mushrooms and continue to cook until the mushrooms begin to soften, about 5 minutes. Add the dashi, ginger, coconut aminos, and monk fruit to the pot. Bring to a simmer, adjust the heat to low, cover, and simmer for 1½ hours, or until the meat is nearly tender (start checking at 10-minute intervals).

Add the sweet potatoes and cook, uncovered, until the stew is thickened and the sweet potatoes are tender, about 20 minutes. Remove from the heat, stir in the lemon juice, and serve garnished with the green onions.

Lamb Tagine with Apricots, Sweet Potatoes, and Artichokes

I went to college in Miami—a cultural melting pot—and some of my best friends there were Middle Eastern. One of my Moroccan friends turned me on to the magic that is the tagine. This extremely savory dish is a really perfect one-pot meal for a cold night. It's a hearty stew with a balance of flavors in every bite—sweet, salty, fresh, and savory—to keep your taste buds excited from start to finish.

Serves 4

2 tablespoons avocado oil
1 large yellow onion, finely diced
2 cloves garlic, minced
1 tablespoon ras-el-hanout*
1 teaspoon ground coriander
1 teaspoon salt, or to taste
1 pound boneless lamb leg meat, cut into
  1-inch cubes
2 cups peeled, diced sweet potatoes
1 cup dried apricots, diced
1 (14-ounce) can quartered artichoke
  hearts, drained and rinsed

1 (15-ounce) can diced tomatoes with juices
2 cups beef bone broth
Zest of 1 lemon
Zest of 1 orange
¼ cup toasted pine nuts
Plain unsweetened coconut yogurt

*Available in most well-stocked spice sections. If unavailable, sub in a mixture of equal parts ground coriander, paprika, and ginger.

Preheat the oven to 375°F.

Heat the oil in a large Dutch oven (or an oven-safe, stovetop-safe tagine, if you have one) over medium-high heat. Add the onion and cook for 3 to 5 minutes, until translucent and tender. Add the garlic, ras-el-hanout, coriander, and salt and stir to combine. Add the lamb and cook until the lamb is browned on the outside and the garlic is fragrant, another 3 to 5 minutes.

Add the sweet potatoes, apricots, artichoke hearts, tomatoes, and bone broth to the pot and bring to a boil. Turn off the heat, cover, transfer to the oven, and cook for 1 hour, or until the lamb is tender. Remove the cover, stir, and cook uncovered for an additional 30 minutes.

Fold in the lemon zest, orange zest, and pine nuts, then taste and adjust the seasoning as needed. Serve with coconut yogurt for creaminess and tang.

Better Brain Bowl

This is a super-simple recipe (if you can even call it that) that provides incredible brain nourishment in the form of monounsaturated fat, lutein, zeaxanthin, omega-3s, and fiber. If you really want to throw caution to the wind and kick things up a notch, add a dollop of chipotle lime avocado oil–based mayo (it can be found at most major supermarkets) after the last step!

Serves 1

1 (4.4-ounce) can sardines (I love Wild Planet Wild Sardines in Extra-Virgin Olive Oil with Lemon)

1 avocado
1 lemon wedge

Drain the oil or water from the can and then empty the sardines into a bowl. Slice up the avocado, add it to the bowl, squeeze the lemon on top, and serve.

Fennel Braised Salmon

In another life, I must have been Greek. These classic Mediterranean flavors are some of my favorites, and you'll see them pop up a few times in this book. If fennel isn't your thing, consider using celery, leek, or even baby bok choy—and adjust the cooking time accordingly.

Serves 4

1 tablespoon extra-virgin olive oil
1 large red onion, thinly sliced
1 bulb fennel, cored and thinly sliced
½ teaspoon salt, or to taste
1 teaspoon ground black pepper
Zest of 1 orange
1 lemon, sliced

½ cup green olives, pitted and roughly chopped
1½ pound skinless salmon fillets, cut into 4 equal pieces
2 cups vegetable broth
1 pinch saffron (optional, and delicious)

Heat the oil in a large high-sided skillet with a tight-fitting lid over medium-high heat. Add the onion, fennel, salt, black pepper, and orange zest and cook, stirring frequently, until the fennel begins to wilt, 5 to 7 minutes. Add the lemon and olives and toss to combine. Add the salmon, vegetable broth, and saffron, if using. Reduce the heat to low, cover, and cook for 10 minutes. Remove the cover and cook for an additional 5 minutes, or until the salmon is tender and flakes easily with a fork. Serve.

Olive Oil–Poached Salmon

This flavorful poached salmon is great on its own, as a topping for salad, or even mixed into a vegetable stir-fry. It's a great recipe for beginners, in that it yields a perfectly tender fish every time—nothing dried out or overly fishy. I recommend using a kitchen thermometer to get the temperature just right, but I have provided instructions just in case you don't have one!

Serves 4 to 6

1 quart extra-virgin olive oil
Zest of 1 lemon, removed in large strips
2 sprigs rosemary
4 cloves garlic, smashed

1 small bunch thyme
1 bay leaf
1 teaspoon salt, or to taste
4 (6-ounce) skinless salmon fillets

Find a pot that will fit all 4 salmon fillets in a single layer and add enough oil to submerge the salmon (leave the salmon out of the pot for now). Add the lemon zest, rosemary, garlic, thyme, bay leaf, and salt to the oil. Heat over low heat until the oil reaches 125°F (the oil should be warm but not simmering/bubbling at all). Gently transfer the salmon to the oil and cook for 20 minutes. Remove the salmon from the oil and serve.

Note: You can strain and keep the oil in the refrigerator to poach another batch of salmon down the road. It will have a slightly fishy taste, but it will also be delicious for stir-fries or sautes, or in salad dressing.

Za'atar Salmon Burgers

This "burger" recipe takes on a tangy, Middle Eastern twist thanks to sumac-heavy za'atar and plenty of citrus zest. I like these patties with a green salad or wrapped in a lettuce leaf as a breadless burger. If you want to really push them over the top (remember: you only live once!), try them with sliced avocado and grilled red onions.

Serves 4

1½ pounds skinless wild salmon fillets
1 shallot, minced
1 tablespoon minced fresh dill
3 tablespoons za'atar*
1 teaspoon salt*
Zest of 1 lemon
Zest of 1 orange
1 tablespoon Dijon mustard

2 tablespoons coconut flour
2 tablespoons almond flour
2 tablespoons extra-virgin olive oil

*Za'atar is a Middle Eastern spice blend that contains sesame seeds, sumac, and other spices. It sometimes also contains salt. If your za'atar contains salt, omit the salt from this recipe.

Cut the salmon into small chunks, then pulse them in a food processor until they are the consistency of ground beef (you can also hand chop the salmon if you'd prefer).

Transfer the salmon to a bowl and fold in the shallot, dill, za'atar, salt, lemon zest, orange zest, mustard, coconut flour, and almond flour until well combined.

Let the mixture rest for 10 minutes so the flour can hydrate, then form into four equal-sized patties. If the mixture doesn't hold together, add additional coconut flour 1 tablespoon at a time and rest briefly before shaping.

In a large skillet, heat the oil over medium heat. Add the patties to the skillet and cook for about 6 minutes per side, until golden brown on the outside and 125°F on the inside.

Serve wrapped in lettuce, on top of a salad, or however you prefer your burgers.

Coconut Curried Eggs

Who says you can't eat eggs for dinner? Not this guy. Whenever I see an egg curry at an Indian restaurant, I order it without thinking twice—because it's such a hearty, savory dish. While many egg curries feature hard-boiled eggs, I drew inspiration from shakshuka to delicately poach eggs in the curry sauce, imparting maximum flavor with each bite.

Serves 4 to 6

2 tablespoons extra-virgin olive oil

1 medium yellow onion, minced

1 medium green bell pepper, minced

2 cups broccoli slaw

2 cloves garlic, minced

3 tablespoons mild curry powder

1 teaspoon paprika (hot or sweet, your choice)

2 teaspoons ground turmeric

1 teaspoon ground cumin

1 teaspoon ground ginger

1 teaspoon salt, or to taste

1 tablespoon monk fruit sweetener

1 (28-ounce) can diced tomatoes

1 (14-ounce) can full-fat unsweetened coconut milk

6 eggs

¼ cup minced fresh basil

Hot sauce

Heat the oil in a very large skillet with a tight-fitting lid over medium heat.

Add the onion, bell pepper, and broccoli slaw and cook for 4 to 5 minutes, until the vegetables are tender. Add the garlic, curry powder, paprika, turmeric, cumin, ginger, salt, and monk fruit and cook until the garlic is fragrant and the spices are lightly toasted, about 2 minutes. Add the tomatoes and coconut milk, bring to a simmer, and simmer for 7 to 10 minutes, until the sauce begins to thicken.

Gently crack the eggs into the pan, reduce the heat to low, cover, and cook until the eggs are set but not tough, 5 to 7 minutes. Sprinkle with the basil and serve with hot sauce.

Brisket Chili, Mole Style

A dish that combines grass-fed beef and dark chocolate? If heaven exists, believe you me, this is being served at the buffet. This beef-based chili draws from classic Tex-Mex flavors as well as the traditional flavors of Mexican mole, made with chiles and chocolate, for a rich, savory flavor that's perfect on a cool winter evening, at a tailgate, or whenever you need a bowl of something warm and cozy.

Serves 6

3 dried ancho chiles
3 dried guajillo chiles
1/4 cup dried mushrooms
1 cup boiling water
1/4 cup avocado oil
2 medium yellow onions, minced
2 pounds beef brisket, diced
1 head garlic, cloves diced
1/4 cup minced fresh cilantro stems
2 tablespoons chili powder
1 tablespoon ground cumin

1 tablespoon dried oregano
1 1/2 teaspoons salt, or to taste
1 (28-ounce) can fire-roasted tomatoes
2 cups beef or chicken bone broth
1/4 cup minced sugar-free bittersweet chocolate (preferably Mexican chocolate)
Plain unsweetened coconut yogurt
1/4 cup minced fresh cilantro leaves
Sliced jalapeños
Pickled onions

Remove the stems and seeds from the chiles, then break into small pieces into a small bowl. Add the dried mushrooms and boiling water and let steep for 10 minutes. Transfer to a blender and blend until smooth. Set aside.

Heat the oil in a large, heavy-bottomed soup pot over medium high heat. Add the onions and brisket and cook until the brisket is browned and the onions are tender, 7 to 10 minutes. Add the garlic and cilantro stems and cook until the garlic is fragrant, 2 to 3 minutes. Add the chili powder, cumin, oregano, and salt and cook for an additional minute. Add the dried chile puree, fire-roasted tomatoes, and bone broth to the mixture, bring to a simmer, then reduce the heat to low and simmer for 1 1/2 to 2 hours, until the beef is fall-apart tender and the chili is nice and thick.

Remove from the heat, add the chocolate, and stir to melt it in. Taste and add salt as needed; if it's too spicy, add more bone broth or coconut yogurt to tone down the heat.

Serve with the cilantro leaves, coconut yogurt, jalapeños, and pickled onions.

Shawarma-Spiced Kebabs

Growing up in New York City, I had access to the best shawarma this side of the Nile. My take on it is that it's a great way to use up the last of the fresh herbs in your fridge—but it's so much more. The mix of dry and fresh herbs and spices really adds a punch of flavor to the meat—and kebabs are great cooked on the stovetop, or even on your grill. I suggest whipping up a quick coconut yogurt sauce for dipping if you've got the time—just combine plain coconut yogurt (or dairy Greek yogurt if your heart desires), herbs, salt, pepper, and lemon juice.

Serves 4

2 tablespoons ground black pepper
2 tablespoons ground allspice
2 tablespoons garlic powder
1 tablespoon ground turmeric
1 tablespoon ground cinnamon
1 tablespoon ground nutmeg
1½ teaspoons ground cloves
1½ teaspoons ground cardamom
1½ teaspoons dried oregano

1½ teaspoons salt
1½ pounds ground lamb*
1 shallot, finely minced
2 tablespoons finely minced fresh mint
2 tablespoons finely minced fresh dill
1 tablespoon extra-virgin olive oil

*You could sub beef if you like.

In a small bowl, whisk together the black pepper, allspice, garlic, turmeric, cinnamon, nutmeg, cloves, cardamom, oregano, and salt.

Place the lamb in a large bowl and toss with 3 tablespoons of the spice mixture (save the remaining spice mixture for another time), then toss with the shallot, mint, and dill. Using clean hands, gently work the lamb mixture until well combined.

Shape into patties (if cooking on stovetop, makes 8 to 10) or around skewers (if grilling, makes 6 to 8). Brush the meat with the oil. Heat a grill to medium heat, sear the meat for 2 to 3 minutes per side. To cook on the stovetop, heat a large skillet to medium-high heat. Add oil, then cook patties 3 to 5 minutes per side, until crisp and brown on the outside and about 130 degrees in the middle. Serve.

Sheet Pan Balsamic Chicken and Broccoli with Red Onions and Figs

Balsamic vinegar and figs are a perfect combination of sweetness and tang—and they pair beautifully with broccoli. I love that this is a single pan meal that's quick, easy, and perfect for a worknight. If you're not a huge fan of chicken or want to change it up, you can make this with salmon or shrimp.

Serves 4

1½ pounds boneless, skinless chicken breasts, cut into 1½-inch strips

1 teaspoon salt, divided

⅓ cup balsamic vinegar, divided

1 teaspoon Dijon mustard

1 large red onion, roughly chopped

4 cups broccoli florets, cut into bite-sized pieces (1 large head)

2 cups peeled and finely diced sweet potato

1 cup chopped dried figs

¼ cup extra-virgin olive oil

2 tablespoons fresh rosemary, minced

2 cloves garlic, minced

½ teaspoon ground black pepper

½ teaspoon dried oregano

Place the chicken in a large bowl and season with ½ teaspoon of the salt. In a small bowl, whisk together half of the vinegar and the mustard and toss with the chicken to coat. Leave the chicken to marinate for 30 minutes to 1 hour.

Preheat the oven to 400°F. Line a large sheet pan with parchment paper.

In a large bowl, toss together the onion, broccoli, sweet potato, and figs. Whisk together the remaining vinegar, remaining ½ teaspoon salt, the oil, rosemary, garlic, black pepper, and oregano and toss with the vegetable mixture.

Transfer the veggies to the prepared sheet pan and bake for 10 to 15 minutes, until just beginning to brown around the edges. Toss the veggies, add the chicken to the pan, and cook for an additional 15 to 20 minutes, until an instant-read thermometer inserted into the thickest part of the chicken reads 165°F and the veggies are cooked to your liking. Let sit for 3 to 5 minutes, then serve.

Roasted Leg of Lamb with Garlic, Sweet Potatoes, and Broccoli

There's nothing I love more than a single-pan (in this case, a roasting pan) dinner for easy cleanup and convenience. And this classic roast lamb is the perfect party dish for anyone who wants to host and still enjoy plenty of time with their guests. The garlicky lamb is offset by the sweetness of citrus zest and pairs perfectly with sweet potatoes and nutty roasted broccoli.

Serves 8 to 10

8 cloves garlic, divided
2 tablespoons fresh rosemary leaves
2 tablespoons fresh thyme leaves
Zest of 1 lemon
Zest of 1 orange
1 teaspoon ground coriander
1 teaspoon ground black pepper

1½ teaspoons salt, divided
1 tablespoon Dijon mustard
1 (5- to 6-pound) bone-in leg of lamb
6 cups broccoli bite-sized florets (1½ large heads)
3 cups peeled and cubed sweet potatoes

Preheat the oven to 325°F. Line a roasting pan with foil (for easier cleanup).

In a food processor, pulse together 4 of the garlic cloves, the rosemary, thyme, lemon and orange zests, coriander, black pepper, 1 teaspoon of the salt, and the mustard to form a smooth paste.

Place the lamb fat-side up in the prepared roasting pan. Using a sharp knife, score the fat by making shallow cuts all over. Rub the garlic paste onto the fatty side of the lamb, coating as much of the exposed surface as possible.

Transfer to the oven and roast until the lamb reaches desired temperature (about 135°F for medium, 120 to 125°F for medium-rare). This will take 1½ to 2 hours, depending on the size of your roast.

Meanwhile, thinly slice the remaining garlic.

Take the lamb from the oven, remove from the pan, tent with foil, and let it rest. Turn the oven temperature up to 400°F.

Immediately put the broccoli, sliced garlic, and sweet potatoes in the roasting pan and toss with the lamb fat in the bottom of the pan. Roast for 15 to 20 minutes, until tender in the middle and crisp on the edges. Serve with the well-rested lamb.

Cocoa-Rubbed Beef Loin with Blueberry–Red Wine Sauce

Time to hire a sitter; this dish, bringing together cocoa, beef, and blueberries, is like a ménage à trois in your mouth. Throw some wine into the mix, and you've got yourself a party! As you might imagine, this dish is rich and indulgent, so it's great to serve at an intimate dinner gathering or for a date night. I've also made the sauce with a mixture of raspberries and blackberries—and it was delicious that way, too, so feel free to work with what you've got on hand.

Serves 4

¼ cup unsweetened cocoa powder
1 tablespoon salt, or to taste
1 teaspoon paprika
1 teaspoon cayenne pepper
½ teaspoon ground cinnamon
½ teaspoon garlic powder
1 (1½-pound) beef tenderloin

4 tablespoons avocado oil, divided
2 shallots, minced
Zest of 1 orange
2 cups blueberries, minced
2 tablespoons fresh thyme leaves
2 cups dry red wine

Preheat the oven to 425°F.

In a small bowl, whisk together the cocoa powder, salt, paprika, cayenne, cinnamon, and garlic powder. Place the tenderloin in a large bowl and toss in the cocoa mixture to coat.

Heat 2 tablespoons of the oil in a very large, oven-safe skillet over medium heat. Sear the tenderloin on all sides, then transfer to the oven. Roast for 30 to 40 minutes, until an instant-read thermometer inserted into the center reads 125°F for rare or up to 150°F for well-done.

Meanwhile, make the sauce: Heat the remaining 2 tablespoons oil in a medium saucepan over medium-high heat. Add the shallots and orange zest and cook until the shallots are tender, then add the blueberries and thyme and cook until the blueberries start to pop. Add the red wine, bring to a simmer, then reduce the heat to maintain a simmer and cook for 20 to 35 minutes, until the liquid has reduced by half and the bulk of the alcohol has cooked off.

When the tenderloin has reached your desired level of doneness, remove from oven, place on a carving board, and let rest 10 minutes. Taste the sauce and add more salt if needed. Slice the meat and serve with the sauce.

Simply Grilled Steak with Charred Kale Salad

Some days, you just need a steak. And whether you're grilling on a big gas grill or a little balcony-sized hibachi, there's no reason to get an extra pan dirty—you can cook your side dish right on the grill, too. I find the char adds a bit of sweetness to the salad, making it a perfect accompaniment to the rich steak.

Serves 2

2 (6-ounce) New York strip steaks
1 teaspoon salt
2 large bunches lacinato kale
4 tablespoons extra-virgin olive oil, divided
1 large red onion, thinly sliced

1 avocado, thinly sliced
¼ cup toasted hazelnuts
¼ cup nutritional yeast
¼ cup balsamic vinegar
1 clove garlic, crushed and minced

Heat half your grill to high heat (if gas) or build a two-zone fire (if charcoal).

Season the steaks with the salt and let come to room temperature while the grill is heating up.

In a large bowl, toss the kale with 2 tablespoons of the oil.

Grill the steaks: for medium-rare, cook for 4 to 5 minutes per side over high heat. For anything more done, transfer to the cooler part of the grill and cook until you've reached your desired temperature—130 to 135°F for medium-rare, 135 to 140°F for medium-well, 150°F for well-done (keep in mind that the steaks will continue to cook a little after you remove them from the grill).

Remove the steaks from the grill and let rest.

Meanwhile, grill the kale over the hottest part of the grill for 30 seconds to 1 minute, just to get a char on the edges. Remove the kale from the grill and, remove leaves from stems (discarding stems), and chop them into bite-sized pieces. Return to the bowl and toss with the remaining 2 tablespoons oil, the onion, avocado, hazelnuts, nutritional yeast, vinegar, and garlic. Serve the steak with the kale salad.

Mushroom Broth with Greens and Ramen-Style Eggs

I'm always game for a hearty and substantial protein-rich soup as a meal, and this recipe does the trick. This ramen-inspired broth is made with mushrooms rather than a pork base, and includes sliced greens for "noodles"—and like any good bowl of ramen, those creamy, rich, perfectly cooked eggs turn dinner from delicious to luxurious.

Serves 4

1 cup dried shiitake mushrooms, crumbled
6 cups water, chicken bone broth, or dashi
2 tablespoons sesame oil
3 cloves garlic, minced
2 tablespoons grated fresh ginger

2 cups fresh mushrooms, diced (assorted types is best)
4 tablespoons coconut aminos or tamari, divided
2 cups sliced hearty greens (cabbage, kale, chard, collard, or mustard greens)
4 eggs

Steep the dried mushrooms in the hot water or broth for 20 minutes, while simmering over low heat.

Meanwhile, in a large, heavy-bottomed soup pot, heat the oil over medium-high heat. Add the garlic, ginger, and fresh mushrooms and cook until the mushrooms are browned and tender, 3 to 6 minutes.

Strain the mushroom broth, pressing on the solids to release all the flavorful liquid.

Add 4 cups of the mushroom broth to the pot, bring to a simmer, then reduce the heat and simmer for 15 minutes.

Pour the remaining 2 cups broth into a separate small pot, add 2 tablespoons of the coconut aminos, and bring to a boil.

While broth is coming to a boil, add the greens and the remaining 2 tablespoons coconut aminos to the large soup mixture and reduce the heat to low. Cover and simmer for 5 to 10 minutes, until the greens are wilted.

When the broth in the small pot is boiling, carefully add the (uncracked) eggs to the pot and cook uncovered for 6 to 7 minutes. Remove from the broth, run under cold water, and peel the eggs.

Slice each egg in half and serve cut-side up in individual serving bowls of broth and greens.

Salmon and Shrimp Stew with Pumpkin and Coconut

This one is inspired by a traditional Thai red curry with pumpkin (or squash) served at some of the amazing Thai restaurants where I live in LA. This hearty recipe is a great stand-alone dinner—and the leftovers make fantastic lunch. Thai red curry paste is available near the soy sauce in most well-stocked grocery stores, and there are a lot of fantastic recipes using it online. It's a little spicy, so if heat isn't your thing, consider adding extra coconut milk or even a spoonful of almond butter or tahini to tone down the heat.

Serves 4

2 tablespoons coconut oil

1 small red onion, diced

4 cups thinly sliced peeled fresh pumpkin or butternut squash

3 cloves garlic, minced

1 heaping tablespoon finely grated fresh ginger

3 tablespoons Thai red curry paste

1 cup pumpkin puree (unsweetened canned pumpkin is perfect)

1 (15-ounce) can full-fat unsweetened coconut milk

2 cups vegetable or chicken bone broth

2 cups peeled and deveined large shrimp

1 cup chopped (bite-sized chunks) skinless salmon

2 cups shredded kale leaves

1 cup frozen peas

Juice of 1 lime

1/4 cup minced fresh cilantro

1/4 cup minced fresh basil

1/4 cup minced fresh mint

1 jalapeño, finely diced (optional; there's already spice from the curry paste)

In a large soup pot, heat the oil over medium-high heat. Add the onion and pumpkin and cook, stirring frequently, until the onion is translucent, 3 to 5 minutes. Add the garlic, ginger, and curry paste and cook for an additional 3 to 4 minutes, until the garlic and ginger are fragrant. Add the pumpkin puree, coconut milk, and broth to the pot and stir well to combine. Reduce the heat to low, cover, and simmer until the pumpkin is tender, about 15 minutes.

Add the shrimp and salmon and continue to simmer for 5 to 7 minutes, until the shrimp is cooked through and the salmon is slightly flaky. Add the kale and peas and cook uncovered for an additional 3 to 5 minutes, until the greens are wilted.

Remove from the heat and add the lime juice, cilantro, basil, mint, and jalapeño, if using, and serve.

Seared Scallops with Fennel Herb Salad and Creamy Almond Sauce

I've always enjoyed scallops, I but rarely order them in restaurants because they're so expensive and you get so few. This dish is perfect if you've got access to nice, big wild-caught scallops. And while I prefer fresh, you can use frozen scallops—just make sure to thaw them thoroughly before starting the dish to get that nice crisp sear. The creamy, tangy romesco-inspired sauce is also great as a dipping sauce for raw veggies—or even thinned out as a salad dressing.

Serves 4

1 cup jarred roasted red bell peppers, drained and patted dry
¾ cup lightly toasted almonds
1 clove garlic, peeled
1 shallot, peeled
1 tablespoon red wine vinegar
1 teaspoon smoked paprika
Zest and juice of 1 lemon
1½ teaspoons salt, divided

¼ cup extra-virgin olive oil
2 bulbs fennel, cored and thinly sliced
1 large red onion, minced
¼ cup minced fresh dill
¼ cup minced fresh parsley
¼ cup minced fresh chives
1 pound sea scallops
¼ cup avocado oil

In a high-speed blender or a food processor fitted with an "S" blade, combine the roasted peppers, almonds, garlic, and shallot and pulse until well combined and chunky. Add the vinegar, paprika, lemon zest, and ¾ teaspoon of the salt. With the machine running, stream in the olive oil through the hole in the lid until a creamy sauce has formed.

In a large bowl, toss together the fennel, onion, dill, parsley, and chives. Season with the lemon juice and a pinch of the remaining salt. Set aside.

Using a paper towel, pat the scallops completely dry on all sides and season with the remaining salt.

Heat a well-seasoned cast-iron skillet over medium-high heat. Add the avocado oil, then the scallops, and cook for 2 minutes, making sure not to move them at all.* Flip and cook for an additional minute. Serve the scallops with the fennel salad and a large dollop of the sauce.

*You may need to work in batches so you don't overcrowd the pan.

Braised Beef and Sweet Potatoes with Cauliflower "Polenta"

Spend any significant time in the Northeast and you know the misery of drudging through cold and rainy weather, desperate to get home after a long day. I've been there, and it was times like those I really appreciated my mother or grandmother making me a good stew. Served over a creamy cauliflower polenta, this take is richly savory, thanks to the addition of miso paste to what would otherwise be a classic Italian recipe.

Serves 4

1 pound beef chuck, cut into bite-sized pieces
½ teaspoon salt, or to taste
1 tablespoon tapioca starch
4 tablespoons avocado oil, divided
1 large yellow onion, thinly sliced
3 ribs celery, diced
3 medium carrots, diced
3 cloves garlic, minced
1 tablespoon red miso paste

1 teaspoon ground black pepper, or to taste
1 teaspoon paprika
1 tablespoon minced fresh rosemary
3 cups beef bone broth
2 cups sweet potatoes, peeled and diced
1 medium head cauliflower, chopped into bite-sized pieces
¼ cup unsweetened coconut cream
¼ cup nutritional yeast

Place the beef in a large bowl, season with the salt, and toss in the tapioca starch.

In a large Dutch oven, heat 2 tablespoons of the oil over high heat. Add the beef and sear until browned on all sides, about 4 minutes per side.

Remove the beef from the pot to a bowl, reduce heat to medium, add the onion, celery, and carrots to the pan, and cook until the onions are translucent and tender, about 5 minutes. Add the garlic, miso paste, black pepper, paprika, and rosemary and cook for 2 to 3 minutes, until very fragrant and well combined. Add the beef broth, bring to a simmer, then reduce the heat to low, cover, and cook until the meat is very tender, about 2 hours. Uncover, add sweet potatoes, and continue to cook, stirring occasionally, for about 30 minutes for the sauce to reduce and thicken.

While the beef has about an hour left to cook, preheat the oven to 400°F.

In a large bowl, toss the cauliflower with the remaining 2 tablespoons oil and spread onto a large sheet pan. Roast the cauliflower until dark golden brown and tender, tossing occasionally, about 40 minutes. Remove the cauliflower from the oven and transfer to a food processor. Add the coconut cream and nutritional yeast and pulse until creamy. Taste and add more salt and pepper if needed.

Place the cauliflower in serving bowls, top with the beef, and serve.

CHAPTER 6

ACCOUTREMENTS

This chapter contains a few sauces that were too good to leave out. I'm a huge fan of sauces, but most commercial preparations (including restaurant offerings) are loaded with sugar and/or unhealthy oils. I've included serving suggestions as well.

Whipped Black Garlic and Olive Oil Dipping Sauce

This sauce is inspired by the addictive garlic sauce (toum) at some of my favorite local Middle Eastern spots, but made using only super healthy oils. If you like a sharper sauce, you can use the traditional raw garlic, but the slightly sweet flavor of fermented garlic is really nice, especially with heartier meats like lamb or beef. Make sure to look for a buttery, mild-flavored extra-virgin olive oil to round out the flavor.

Makes 1 cup, serves 8

¼ cup avocado oil

1 shallot, thinly sliced

1 teaspoon salt

2 bulbs black garlic, peeled

⅓ cup mild-flavored extra-virgin olive oil

⅓ cup lemon juice

In a small saucepan, heat the avocado oil over low heat. Add the shallot and cook 3 to 4 minutes, until tender. Set aside to cool. Transfer the shallot with the oil to a food processor fitted with an "S" blade and add the salt and black garlic.

With the machine running, stream in 1 tablespoon of the olive oil, then 1 tablespoon lemon juice, scraping down the sides as needed. When the oil and lemon juice incorporate and emulsify, add another round until you've used it all, continuing to scrape the sides as needed. This process will take up to 15 minutes and the result will be a light, fluffy sauce that's delicious as a dip for raw veggies or roasted chicken or beef.

Store in the fridge, covered and wrapped well (the strong flavor of black garlic likes to spread).

Savory Blueberry Balsamic Sauce

As I've mentioned before, I love balsamic vinegar, and I relish any opportunity to use it in recipes. This sauce is deceptively simple to make, but it's nonetheless incredibly rich, complicated, tangy, sweet, and super-savory. It goes perfectly with grass-fed beef or pork or a strong-flavored fish like salmon. You can also use it for a dipping sauce or drizzle it on roasted veggies.

Makes about 1 cup

2 tablespoons extra-virgin olive oil
2 shallots, minced
1 tablespoon fresh thyme leaves
1 teaspoon ground black pepper

¼ teaspoon salt
1 heaping cup fresh or frozen blueberries
¾ cup balsamic vinegar

Heat the oil in a small saucepan over medium-high heat. Add the shallots, thyme, black pepper, and salt and cook for 3 to 5 minutes, stirring frequently, until shallots are tender. Add the blueberries and cook until the blueberries burst. Add the vinegar, reduce the heat to low, and cook for 10 to 15 minutes, thoroughly crushing the berries with a fork or whisk until berries have released their juices and a sauce has formed.

Transfer to a blender and blend until smooth. Strain the sauce through a fine-mesh strainer back into the pan. Place over low heat, bring to a simmer, and cook until the mixture is reduced to 1 cup. Store refrigerated in an airtight glass jar for up to 2 weeks.

Almond "Peanut" Butter

I like peanut butter as much as the next guy, but for a lot of people, peanuts are off the table. Whether you're someone who is in the "no peanut" crew, or you're just looking for a slightly healthier alternative . . . this almond butter delivers. It's got a rich, peanutty taste despite the absence of peanuts and is great with jam, dark chocolate, or even out of the jar. Just know that once you've broken the seal, it may be hard to stop! Don't say I didn't warn you. Use anywhere you'd use peanut butter.

Makes about 2 cups

2½ cups raw almonds

½ cup raw macadamia nuts (optional, or sub another ½ cup almonds)

1½ teaspoons coconut aminos

2 tablespoons nutritional yeast

½ teaspoon monk fruit sweetener

Heat a large skillet over medium-low heat. Add the almonds and macadamia nuts and toast, stirring constantly, until they smell fragrant and nutty, taking care not to let them burn, about 6 minutes. Transfer to a plate and let cool to room temperature.

Transfer the nuts to a food processor fitted with an "S" blade and pulse until gritty, then process the nuts until a paste is formed (this can take 5 minutes or more, depending on your machine). Make sure to scrape down the sides regularly.

When nut butter is *almost* to the consistency of your liking (smooth or chunky—your choice), add the coconut aminos, nutritional yeast, and monk fruit and process for 1 minute more, or until you've reached the perfect consistency.

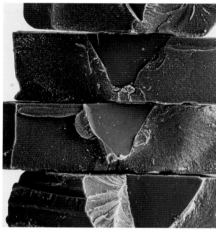

CHAPTER 7

DESSERTS

Who says living a Genius Life can't include dessert? I certainly don't! In this chapter, you'll find sweet treats made without grain flours or sugar so you can enjoy indulging and feel great afterward. It's possible, and these recipes are proof!

Almond Olive Oil Cake

Olive oil cakes are traditional in Italian cuisine—though they're often made with grain flour rather than almonds. This luscious, protein-rich cake highlights some flavors that are a bit unusual for a dessert—rosemary and black pepper—and which in fact go beautifully with the citrus, cranberry, and allspice. To push it over the top, try it with a scoop of Sugar-Free Olive Oil Ice Cream (page 264).

Serves 8 to 12

1/3 cup extra-virgin olive oil, plus more for the pan

Zest of 1 orange, cut into large strips

1 sprig rosemary

1 teaspoon whole peppercorns

1¾ cups almond flour (blanched is best)

1 tablespoon coconut flour

1¼ cups monk fruit sweetener

1½ teaspoons salt

½ teaspoon baking soda

½ teaspoon baking powder

¾ teaspoon ground allspice

1¼ cups full-fat canned unsweetened coconut milk

3 large eggs

1 teaspoon almond extract

¼ cup orange juice (from the zested orange, above)

1 cup fresh, frozen, or dried unsweetened cranberries

Preheat the oven to 350°F. Spray a 9-inch round springform cake pan at least 2 inches high with oil and line it with parchment. Set aside.

In a small saucepan, combine the oil, orange zest, rosemary, and peppercorns and bring to a simmer over low heat. Simmer for 10 minutes, then remove from the heat and let cool.

In a large bowl, whisk together almond flour, coconut flour, monk fruit, salt, baking soda, baking powder, and allspice. In a separate bowl, whisk together coconut milk, eggs, almond extract, and orange juice.

Strain the cooled oil and add it to the egg mixture. Add the wet ingredients to the dry and whisk until just combined. Gently fold in the cranberries.

Pour into the prepared cake pan and bake for 45 minutes. Cover with foil and bake for an additional 20 to 30 minutes, until the cake is golden brown and a toothpick inserted into the center comes out clean.

Allow to cool on a wire rack in the pan for at least 30 minutes, then remove the cake by running a knife around the edges and carefully removing the ring of the pan. Slice and serve.

Avocado Chocolate Truffles

If you've got a sweet tooth, this one's for you. You'll really want to use a food processor for these so you have a perfectly smooth base for your truffles (chunks of unpureed avocado would be a buzzkill). You can try folding in some finely chopped toasted nuts or even coconut to tweak the flavor or up the texture of these rich, delicious, very dark chocolate truffles.

Makes 18 to 20

1 avocado (make sure it's very ripe)

1 cup bittersweet chocolate chips (at least 72 percent cacao), melted in double boiler

½ teaspoon pure vanilla extract

¼ teaspoon salt

Liquid stevia (3 to 6 drops are likely enough)*

¼ cup unsweetened cocoa powder

*I use unflavored stevia, but mint- or orange-flavored drops also complement the flavor of these truffles.

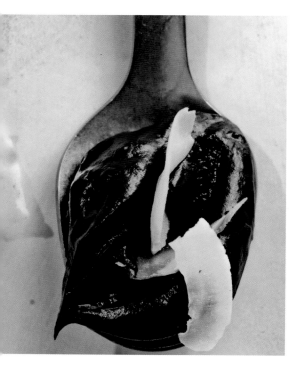

In a food processor fitted with an "S" blade, process the avocado until smooth, scraping down the sides a few times. Add the melted chocolate, vanilla, and salt and process until smooth, scraping the sides frequently. Taste and add stevia 1 to 2 drops at a time until the mixture is as sweet as you like. Transfer to a container and refrigerate for at least 1 hour to get it to a scoopable consistency.

Use a small spring-loaded ice cream scoop or a tablespoon to separate the mixture into truffle-sized lumps. Working quickly, roll each lump into a smooth ball with your hands (this can get messy—wear disposable gloves if desired). Spread the cocoa powder out on a plate and toss the truffles in the cocoa powder.

Serve at room temperature or slightly chilled. Store leftovers in the refrigerator for up to 2 weeks.

Actually Healthy Hot Chocolate

On a winter day, there's nothing like a warm, comforting cup of hot chocolate—and this, in my opinion, is the ultimate recipe. Not only is it just about the richest hot cocoa recipe out there, but it reminds me of one of those chocolate oranges Hannukah Harry would bring me as a kid. (Fun fact: Hannukah Harry actually traces back to a 1989 sketch on *Saturday Night Live*. My family appreciated finally having a gift-giving patron saint for our own holiday, even if he was really just Jon Lovitz in costume.) This hot chocolate includes MCT oil, a fraction of coconut oil, which has been shown to support brain energy needs—how cool is that?

Serves 2

2 cups canned unsweetened coconut milk

1 cup water

¼ cup bittersweet chocolate chips (at least 72 percent cacao)

1 teaspoon cocoa powder

Zest of 1 orange

1 teaspoon monk fruit sweetener, plus more as needed

¼ teaspoon salt

1 teaspoon pure vanilla extract

2 tablespoons MCT oil

Combine the coconut milk and water in a large saucepan over medium heat and bring to a simmer. Reduce the heat to very low, add the chocolate chips, cocoa powder, orange zest, monk fruit, and salt, and cook, stirring constantly, until the chocolate chips are melted and the monk fruit is dissolved.

Remove from the heat, stir in the vanilla and MCT oil, and serve immediately.

Sugar-Free Olive Oil Ice Cream

This simple dairy-free ice cream recipe is inspired by the incredible olive oil ice cream at Gelateria Uli in Los Angeles, with the addition of rosemary, because rosemary and olive oil pair absolutely beautifully in desserts (you'll also find them in the olive oil cake on page 260). Make sure to use unsweetened coconut cream for this dessert (often found among Asian ingredients in the grocery store) rather than sugar-heavy cream of coconut.

Makes just under 1 quart

2 (14-ounce) cans unsweetened coconut cream
1 sprig rosemary
½ cup monk fruit sweetener
1 scant pinch salt
1 vanilla bean pod
6 egg yolks
½ cup very-high-quality extra-virgin olive oil

1 tablespoon vodka (optional)*

*The vodka helps the ice cream stay scoopable when freezing and won't impact the flavor of the final product. It's fine to make your ice cream without it, but you'll want to let the ice cream sit at room temperature for 10 to 15 minutes before serving or it will be impossible to scoop.

In a large saucepan, combine the coconut cream, rosemary, monk fruit, and salt and bring to a simmer over medium heat. Split the vanilla bean in half and scrape the contents into the coconut mixture, saving the bean for a different use (like making your own vanilla extract). Simmer, stirring occasionally, for 10 to 15 minutes, until the monk fruit is dissolved. Set aside to cool, then remove the rosemary from the coconut mixture and discard.

Place the egg yolks in the bowl of stand mixer and whisk until frothy. With the mixer running, carefully and slowly stream the warm coconut mixture into the eggs, taking care that the eggs don't curdle. Transfer back to the saucepan and cook over low heat, stirring constantly, until the mixture thickens enough to coat a wooden spoon, 3 to 4 minutes.

Strain through a fine-mesh strainer and whisk in the olive oil and vodka, if using. Transfer to the refrigerator and chill completely, at least 1 hour or as long as overnight. This is a great time to make sure your ice cream machine components are chilled, too.

Freeze according to your ice cream maker's instructions to the consistency of soft serve. If you have any remaining olive oil, drizzle a small amount over the top (optional). Serve immediately, or transfer to an airtight container and freeze to a hard-scoop consistency.

Note: If you don't have an ice cream machine, pour the chilled ice cream base into a freezer-safe loaf pan. Remove from the freezer every 20 to 30 minutes and stir, making sure to thoroughly scrape all the edges. This is a long (2- to 3-hour) process and will result in a slightly grainier (but 100 percent delicious) ice cream than you'll get using an ice cream machine.

Chocolate Almond Smoothie

This smoothie is basically a milkshake—but, like, healthy. The frozen banana adds some much-needed sweetness to unsweetened cocoa powder, and the cloves really bring out the flavor of the chocolate and almonds. So, if you're someone who craves a milkshake—try this instead (with a grass-fed burger, if you want). You can add a serving of vanilla whey protein isolate or your protein powder of choice to up the protein—perfect for a post-workout shake.

Serves 1

$\frac{1}{2}$ frozen banana, broken into chunks

1 cup unsweetened plain almond milk, plus more as needed

3 tablespoons unsweetened almond butter

$\frac{1}{2}$ teaspoon almond extract

2 tablespoons unsweetened cocoa powder

$\frac{1}{8}$ teaspoon ground cloves

Monk fruit sweetener, if needed

In a blender, combine the banana, $\frac{1}{2}$ cup of the almond milk, and the almond butter and blend until creamy. Add the remaining $\frac{1}{2}$ cup almond milk, the almond extract, cocoa powder, and cloves and blend until smooth. If needed, add more almond milk a tiny bit at a time until you've reached your desired consistency.

Taste, and if it's not sweet enough, blend in monk fruit sweetener $\frac{1}{4}$ teaspoon at a time. Serve immediately.

Chocolate Coconut Balls

While a little higher on the carbohydrate spectrum, this sweet snack exploits the delicious properties of Medjool dates (I could seriously eat those like they're going out of style). Not only does it satisfy the sweet tooth perfectly, but it's rich in polyphenols, minerals, vitamins, and healthy fats, plus a small hit of protein from the almonds. They're great as a dessert, or for a peri- or post-workout pick-me-up, you can pop one of these "cookie balls" and feel good about it—as well as a bit indulgent.

Makes 12

½ cup raw almonds

1½ cups finely shredded unsweetened coconut, divided

1½ cups pitted Medjool dates (about 15)

3 tablespoons unsweetened cocoa powder

1 tablespoon ground flaxseeds

3 tablespoons toasted sesame seeds

¼ teaspoon salt

In a food processor fitted with an "S" blade, combine the almonds and 1 cup of the coconut and process until a fine crumble is formed, scraping down the sides as needed. Add the dates, cocoa powder, and flaxseeds and continue to process until a smooth paste is formed. You'll need to continue to scrape down the sides frequently and be patient—this process will take a between 3 and 10 minutes, depending on how powerful your food processor is. Add the sesame seeds and salt and pulse until just combined but still a little crunchy. Transfer the mixture to a large bowl and pour the remaining ½ cup shredded coconut onto a plate.

With damp hands, roll the mixture into 12 equal-sized balls, then roll the balls in the shredded coconut. Store in the refrigerator for up to 2 weeks and serve as a quick one- or two-bite snack.

Dairy-Free Blueberry Ice Cream

There's nothing like a big scoop of ice cream on a hot day . . . especially when the ice cream is sweet, tangy, and full of flavor. Thanks to the high (healthy) fat content in this dairy-free ice cream, you'll enjoy the same richness and creaminess as dairy ice cream, whether you eat it soft serve (right out of the ice cream machine) or let it firm up in the freezer until scoopable. The sweetness comes only from blueberries, so you'll feel charged up with phytonutrients—not crashing from sugar—afterward.

Serves 4 to 6

2 cups fresh or frozen blueberries
Zest of 1/2 orange
Zest and juice of 1 lemon
2/3 cup monk fruit sweetener
1/4 teaspoon ground allspice
1 teaspoon pure vanilla extract
1 (14-ounce) can full-fat unsweetened coconut milk

1 (14-ounce) can unsweetened coconut cream
1/4 teaspoon salt
1 tablespoon vodka (optional)*

*The vodka helps keep the ice cream from freezing too solid. If you don't add it, leave the ice cream out at room temperature for 10 to 15 minutes before scooping.

In a saucepan with lid, combine the blueberries, orange zest, lemon zest and juice, monk fruit, allspice, and vanilla and heat over low heat until the blueberries begin to break down, stirring frequently to prevent burning. Remove from the heat and let the mixture cool, then transfer to a blender along with the coconut milk, coconut cream, salt, and vodka, if using. Blend for 2 to 3 minutes, until completely smooth and creamy—the mixture will separate at first and then come back together.

Transfer to a container, place in the refrigerator, and chill completely, at least 1 hour or as long as overnight. This is a great time to make sure your ice cream machine components are chilled. Freeze according to your ice cream maker's instructions to the consistency of soft serve.

Serve immediately, or transfer to an airtight container and freeze to a hard-scoop consistency.

Note: If you don't have an ice cream machine, pour the chilled ice cream base into a freezer-safe loaf pan. Remove from the freezer every 20 to 30 minutes and stir, making sure to thoroughly scrape all the edges. This is a long (2- to 3-hour) process and will result in a slightly grainier (but 100 percent delicious) ice cream than you'll get using an ice cream machine.

Chocolate Blueberry Clusters

If you think chocolate-covered strawberries are delicious, just wait until you try chocolate-covered blueberries! They're as tangy and delicious but easier to pop a cluster and enjoy. So quick, easy, and fun to put together and you can make the clusters as big or small as you want. You can add a hit of lion's mane mushroom powder for an additional brain boost.

Serves 6

1 cup bittersweet chocolate chips (at least 72 percent cacao)*

1 tablespoon coconut oil

1 tablespoon lion's mane extract (optional)

1 tablespoon monk fruit sweetener (optional)**

2 cups fresh blueberries (make sure they are completely dry)

¼ teaspoon fine salt

*You can use two broken-up 3-ounce chocolate bars if you don't have chocolate chips.

**If using a darker chocolate (85 percent or higher), I recommend adding an additional 1 tablespoon monk fruit before folding in the blueberries.

Line a sheet pan with parchment paper and make sure there's room to fit it level in your refrigerator.

Melt the chocolate with the oil over a double boiler, or in 10-second bursts in a pot over low heat, stirring as you go until completely smooth and creamy. Add the lion's mane extract, if using. Remove from the heat. If you used a higher cacao bar (85 percent or higher), stir in the monk fruit. Fold in the blueberries.

While the chocolate is still moldable, use a spoon to pile the blueberries in bite-sized clusters on the sheet pan, 3 to 5 berries per cluster. With a rubber spatula, scrape any remaining chocolate over the berry clusters, making sure they're evenly coated Sprinkle with the salt and transfer to the refrigerator. Let chill for at least 20 minutes, until firmly set. Store extras in the refrigerator for up to a week until you're ready to eat them.

Chocolate Coconut Cookies with Dried Blueberry "Chips"

There's a reason chocolate chip cookies are snack-time staple. They're just so good! This version is gluten- and grain-free, with nutrient-dense ingredients so tasty you won't know you're actually eating something healthy. I love the tart touch the blueberries add, but if you'd prefer (and sometimes I do), you can leave them out—or replace them with unsweetened dried cranberries or chopped dried figs. Just make sure your fruit of choice doesn't have any added sugar.

Makes 12 to 15

¼ cup plus 3 tablespoons coconut oil
¼ cup monk fruit sweetener
½ teaspoon salt
½ teaspoon ground cinnamon
4 large egg yolks
¼ cup almond flour
2 tablespoons arrowroot powder

1 tablespoon ground flaxseeds
½ teaspoon pure vanilla extract
1 cup unsweetened coconut flakes
1 cup bittersweet chocolate chips (or chopped chocolate)—at least 70 percent cacao
¾ cup dried unsweetened blueberries

Preheat the oven to 350°F. Line a sheet pan with parchment paper or a silicone mat.

In a stand mixer fitted with a paddle attachment or a large bowl, beat together the oil, monk fruit, salt, cinnamon, and egg yolks until smooth and fluffy. Add the almond flour, arrowroot, flaxseeds, vanilla, and coconut flakes and beat until a cohesive dough is formed. Fold in the chocolate chips and blueberries.

Roll the cookie dough into a 2-inch diameter log, wrap in plastic, and refrigerate for at least 30 minutes (or as long as overnight). Slice the log into 1-inch-thick cookie shapes and press lightly to spread each out to a thickness of ½ inch to ¾ inch. Bake for 15 minutes, or until the cookies are golden brown at the edges and hold together. Let cool completely on the sheet before serving. Store at room temperature in an airtight container for 3 to 5 days.

Chocolate Avocado Pudding Three Ways

When I was a kid it was always exciting to get a pudding cup in my school lunch. And you know what? I still think pudding is delicious, but it's hard to find a healthy alternative. Never fear; Max is here with a recipe so good and healthy, you'll soon be "puddin'" this creamy, avocado-based recipe among your favorites. You can eat this as is for a twist on a classic chocolate pudding, or try any of my three variations below to spice things up, take a tropical taste bud vacation, or enjoy a rich, indulgent, nutty dessert.

Serves 4

For the base:

2 avocados (make sure they're very ripe), flesh scooped out

1/4 cup powdered Swerve sweetener, plus more as needed

1/4 cup unsweetened cocoa powder

1/4 teaspoon instant espresso powder

2 tablespoons full-fat canned coconut milk, plus more as needed

1 teaspoon pure vanilla extract

1/8 teaspoon salt

FOR A BASIC PUDDING:

In a food processor fitted with an "S" blade, combine the avocados, Swerve, and cocoa powder and process until well combined. Add the espresso powder, 2 tablespoons of the coconut milk, the vanilla, and salt and process until smooth, scraping the sides of the machine a few times. Taste and add more Swerve if you like. Serve as is or proceed to one of the variations below.

Spiced Chocolate Pudding

1/4 cup unsweetened coconut cream

1/4 teaspoon ground cinnamon

1/4 teaspoon ground cloves

1/4 teaspoon cayenne pepper

1/8 teaspoon ground cumin

In a small saucepan, combine all the ingredients, bring to a simmer over medium heat, and simmer for 5 minutes (this can be done up to 1 day in advance). Let cool completely. Add to the basic pudding still in the food processor and process until incorporated. Transfer to a container and chill a bit before serving or it will be slightly too runny.

Coconutty Chocolate Pudding

¼ cup unsweetened shredded coconut

¼ teaspoon coconut extract

¼ cup toasted unsweetened coconut chips

Add the shredded coconut and coconut extract to the basic pudding still in the food processor and process until incorporated (it may still be a little chunky—that's fine). Scoop into individual serving glasses/bowls, top with the toasted coconut chips, and serve.

Toasted Nut Chocolate Pudding

¼ teaspoon almond extract

¼ cup unsweetened coconut cream

2 tablespoons minced toasted pecans

2 tablespoons minced toasted pistachios

2 tablespoons minced toasted walnuts

Add the almond extract to the food processor when making the basic pudding and process with the rest of the ingredients.

In a small saucepan, combine the coconut cream with half of the toasted nuts, bring to a simmer over low heat, and simmer for 10 to 15 minutes to infuse the coconut cream with the flavor of the nuts. Remove from the heat and cool completely.

Blend the nut mixture in with the basic pudding still in the food processor and process until incorporated. Scoop into individual serving glasses/bowls and serve topped with the remaining nuts.

No-Sugar Almond "Granola"

I consider granola more of a dessert than a meal, but there's no doubt that it makes a tasty cereal or topping for coconut or Greek yogurt. The trouble with most super-market versions is that they're loaded with sugar and unhealthy oils. If you can't find the monk fruit "maple syrup" I've used, make a simple syrup by dissolving one part monk fruit in one part water. It won't have quite the same maple flavor, but it still works.

Makes 3½ cups

Avocado oil spray, for the pan
¾ cup unsweetened coconut flakes
2 cups raw slivered almonds
2 tablespoons ground flaxseeds
¼ cup whole flaxseeds
¼ cup toasted sesame seeds
¼ cup unsweetened dried blueberries
½ teaspoon ground cinnamon

½ teaspoon ground ginger
⅛ teaspoon ground cloves
¼ teaspoon salt
¼ cup Lakanto monk fruit maple-flavored syrup
1 teaspoon pure vanilla extract
1 egg white, whipped until frothy

Preheat the oven to 300°F. Line a sheet pan with parchment paper and grease it lightly with avocado oil.

In a large bowl, toss together the coconut flakes, almonds, ground and whole flax-seeds, sesame seeds, and blueberries.

In a small bowl, whisk together the cinnamon, ginger, cloves, and salt.

Add the maple-flavored syrup, vanilla, and egg white to the nut mixture and stir un-til well combined. Add the spice mixture and stir until well incorporated.

Spread the mixture onto the sheet pan and press to form a thin, even layer. Bake for 15 to 20 minutes, until lightly browned. Let cool completely before breaking it into pieces. Store at room temperature.

Chocolate Almond Butter Cups

True story: Growing up, I once ate so many Reese's Peanut Butter Cups in one sitting that I made myself sick, and then for years afterward I thought I was allergic to the combination of peanut butter and chocolate. Thankfully, I was wrong—I just had caved in to the delicious and super-addictive powers of nut butter and chocolate. These almond butter cups are much better for you than commercially sold cups and have a slightly peanutty taste thanks to the umami-rich coconut aminos that go into the spiced almond butter mix.

Makes 24

½ cup smooth unsweetened almond butter
¼ teaspoon ground cinnamon
¼ teaspoon ground cloves
1 to 2 drops liquid stevia
¼ teaspoon coconut aminos
¼ teaspoon salt

1 tablespoon ground flaxseeds
2 cups bittersweet chocolate chips (or chopped bittersweet chocolate)
1½ tablespoons coconut oil
Flaky sea salt, such as Maldon, to garnish

Line a 24-cup mini-muffin tin with paper liners.

In a medium bowl, whisk together the almond butter, cinnamon, cloves, stevia, coconut aminos, salt, and flaxseeds until well combined. Cover and chill in the refrigerator until thickened, 10 to 15 minutes.

Remove from the refrigerator and roll into a log slightly smaller in diameter than the cups in the muffin tin. Cut the log into 24 equal-sized pieces and return to the refrigerator.

Melt together the chocolate chips and oil over a double boiler, stirring until smooth.

Drop a scant teaspoon of melted chocolate into each of the prepared cups and shake or tap the pan to level the chocolate layer. Transfer to refrigerator to chill and set for 5 to 10 minutes.

Place the discs of filling on top of the chocolate and top with remaining chocolate, making sure each disk is thoroughly covered. Tap to eliminate air bubbles.

Sprinkle with salt and chill until set, about 30 minutes. Bring back to room temperature before serving.

Flourless Sugar-Free Blueberry Lemon Tart with Almond Crust

In my opinion, lemon bars are one of the most perfect desserts in the world: it's the ideal balance of tangy and sweet. But how can you make a perfect lemon dessert without the grain, sugar, or dairy? Actually, it's pretty easy and truly delicious, especially with the addition of blueberries to add a little "pop" to the tang of the lemon filling.

For the crust:

1 large egg

2 tablespoons powdered Swerve sweetener

3 tablespoons coconut oil, melted and cooled, plus more if needed

1½ cups almond flour

½ cup plus 2 tablespoons arrowroot powder

1 tablespoon coconut flour

¼ teaspoon ground cloves

¼ teaspoon salt

For the filling:

Zest of 1 lemon

Zest of 1 tangerine

¾ cup fresh lemon juice

4 large eggs

⅓ cup powdered Swerve sweetener

⅓ cup unsweetened coconut cream

1½ cups fresh blueberries

To make the crust: Preheat the oven to 375°F.

In a large bowl, whisk together the egg, Swerve, and coconut oil until well combined.

In a separate bowl, combine the almond flour, arrowroot, coconut flour, cloves, and salt.

Add the wet ingredients to the dry ingredients and stir until well incorporated and easy to handle. If the dough is too dry, add a bit more coconut oil; if it's too wet, add more almond flour.

With damp fingers, pat the crust into a 9-inch tart pan and put the pan on a sheet tray (for easy handling). Bake for 10 to 12 minutes, until golden brown at edges, then remove from the oven and reduce the oven temperature to 325°F while you cool the crust.

To make the filling: In a large bowl, whisk together the lemon and tangerine zests, the lemon juice, eggs, Swerve, and coconut cream until smooth (you can also use a blender).

Spread the blueberries over the baked tart in a single layer. Carefully pour the lemon filling around the blueberries, being careful not to overfill the tart (the filling should come just below the edges of the tart shell).

Return to the oven and bake for 25 to 35 minutes, until the filling is set but just a little jiggly in the center. Let cool completely before serving. Store in an airtight container in the refrigerator for 3 to 5 days.

ACKNOWLEDGMENTS

I loved creating this cookbook for you, and I owe a debt of gratitude to my little team who made it more fun than I could have ever imagined. First to my agent, Giles Anderson, who has believed in my vision of a healthier world since day one, and to my publisher, Karen Rinaldi, who has made this work possible.

Thank you to Kate Holzhauer for helping to bring these recipes to life.

The beautiful photos in this book would not be possible without the talents of photographer Eric Wolfinger and food stylist Alison Attenborough.

Thank you to the cities of New York, Miami, and Los Angeles for the food inspiration.

Huge thanks to my assistant, Sydney Clarke, who's basically like my hired boss and runs all things The Genius Life. You are the best!

Thank you to my family: Andrew, Ben, Bruce, Delilah the cat.

Eternal gratitude to my mother, Kathy, the inspiration for everything that I do. I wish everyone could have a mother like her. I love and miss you, Mom.

Thank you to every social media follower, every podcast listener, every newsletter subscriber . . . I'm grateful you've chosen to give your attention to me.

RESOURCES

Science continues to evolve, and our food landscape is always changing. Keep up with me on social media for additional recipes and to continue your journey of self-education. And listen to my podcast, with new episodes released weekly. I routinely interview experts covering topics such as brain health, hormone health, mental health, fat loss and fitness, and productivity, and offer exclusive discounts and even freebies from partner food brands that I trust.

Find me on social media:
Instagram: @maxlugavere
Twitter: @maxlugavere

Listen to my podcast: http://maxl.ug/tgl or search for "The Genius Life" in your podcast app of choice

Subscribe to my weekly science newsletter: http://maxlugavere.com

Visit my resources page for exclusive discounts from brands I trust: http://maxl.ug/GKresources

NOTES

1. Max Lugavere, Alon Seifan, and Richard S. Isaacson, "Prevention of Cognitive Decline," *Handbook on the Neuropsychology of Aging and Dementia*, ed. Lisa D. Radvin and Heather L. Katzen (Springer, 2019), 205–29; "Non-Pharmacological Approaches to Managing Patients with Alzheimer's Disease," AcademicCME, retrieved March 5, 2021, from https://academiccme.com/courses/non-pharmacological-approaches-to-managing-patients-with-alzheimers-disease/.
2. Bamini Gopinath et al., "Association between Carbohydrate Nutrition and Successful Aging Over 10 Years," *The Journals of Gerontology, Series A,* 71,10 (2016): 1335–40, doi:10.1093/gerona/glw091; Maryam S. Farvid et al., "Fiber Consumption and Breast Cancer Incidence: A Systematic Review and Meta-Analysis of Prospective Studies," *Cancer* 126,13 (2020): 3061–75, doi:10.1002/cncr.32816.
3. Guillermo Cásedas et al., "The Metabolite Urolithin-A Ameliorates Oxidative Stress in Neuro-2a Cells, Becoming a Potential Neuroprotective Agent," *Antioxidants* (Basel, Switzerland) 9,2 (2020): 177, doi:10.3390/antiox9020177.
4. Cristian Del Bo' et al., "Systematic Review on Polyphenol Intake and Health Outcomes: Is There Sufficient Evidence to Define a Health-Promoting Polyphenol-Rich Dietary Pattern?," *Nutrients* 11,6 (2019): 1355, doi:10.3390/nu11061355.
5. R. Dashwood et al., "Study of the Forces of Stabilizing Complexes between Chlorophylls and Heterocyclic Amine Mutagens," *Environmental and Molecular Mutagenesis* 27,3 (1996): 211–18, doi:10.1002/(SICI)1098-2280(1996)27:3<211::AID-EM6>3.0.CO;2-H.
6. Banafshe Hosseini et al., "Effects of Fruit and Vegetable Consumption on Inflammatory Biomarkers and Immune Cell Populations: A Systematic Literature Review and Meta-Analysis," *American Journal of Clinical Nutrition,* 108,1 (2018): 136–55, https://doi.org/10.1093/ajcn/nqy082; Taylor C. Wallace et al., "Fruits, Vegetables, and Health: A Comprehensive Narrative, Umbrella Review of the Science and Recommendations for Enhanced Public Policy to Improve Intake," *Critical Reviews in Food Science and Nutrition,* 60,13 (2020): 2174-11, doi:10.1080/10408398.2019.1632258.
7. Bob Fischer and Andy Lamey, "Field Deaths in Plant Agriculture," *Journal of Agricultural and Environmental Ethics* 31,4 (2018): 409–28, doi:10.1007/s10806-018-9733-8.

8. Morgan E. Levine et al., "Low Protein Intake Is Associated with a Major Reduction in IGF-1, Cancer, and Overall Mortality in the 65 and Younger but Not Older Population," *Cell Metabolism* 19,3 (2014): 407–17, doi:10.1016/j.cmet.2014.02.006.

9. Alan A. Aragon and Brad Jon Schoenfeld, "How Much Protein Can the Body Use in a Single Meal for Muscle-Building? Implications for Daily Protein Distribution." *Journal of the International Society of Sports Nutrition* 15, 10 (2018). doi: 10.1186/s12970-018-0215-1; David Rogerson, "Vegan Diets: Practical Advice for Athletes and Exercisers," *Journal of the International Society of Sports Nutrition* 14, 1 (2017): 1–15.

10. Stuart Phillips et al. "Commonly Consumed Protein Foods Contribute to Nutrient Intake, Diet Quality, and Nutrient Adequacy," *The American Journal of Clinical Nutrition* 101, 6 (2015): 1346S–1352S.

11. Nicholas V. C. Ralston and Laura J. Raymond, "Mercury's Neurotoxicity Is Characterized by Its Disruption of Selenium Biochemistry," *Biochimica et Biophysica Acta* 1862.11 (2018), doi:10.1016/j.bbagen.2018.05.009.

12. Ondine van de Rest et al., "*APOE* ε4 and the Associations of Seafood and Long-Chain Omega-3 Fatty Acids with Cognitive Decline," *Neurology* 86.22 (2016): 2063–70, https://doi.org/10.1212/WNL.0000000000002719.

13. Maria A. I. Åberg et al., "Fish Intake of Swedish Male Adolescents Is a Predictor of Cognitive Performance," *Acta Paediatrica* 98.3 (2009): 555–60, https://doi.org/10.1111/j.1651-2227.2008.01103.x; Jianghong Liu et al., "The Mediating Role of Sleep in the Fish Consumption–Cognitive Functioning Relationship: A Cohort Study," *Scientific Reports* 7.1, article 17961 (2017).

14. Joseph R. Hibbeln et al., "Relationships between Seafood Consumption during Pregnancy and Childhood and Neurocognitive Development: Two Systematic Reviews," *Prostaglandins, Leukotrienes and Essential Fatty Acids* 151 (2019): 14–36, doi:10.1016/j.plefa.2019.10.002.

15. Aesun Shin et al., "Dietary Mushroom Intake and the Risk of Breast Cancer Based on Hormone Receptor Status," *Nutrition and Cancer* 62,4 (2010): 476–83, doi:10.1080/01635580903441212; Shu Zhang et al., "Mushroom Consumption and Incident Dementia in Elderly Japanese: The Ohsaki Cohort 2006 Study," *Journal of the American Geriatrics Society* 65,7 (2017): 1462–69, doi:10.1111/jgs.14812.

16. Katie R. Hirsch et al., "Cordyceps Militaris Improves Tolerance to High-Intensity Exercise after Acute and Chronic Supplementation," *Journal of Dietary Supplements* 14,1 (2017): 42–53, doi:10.1080/19390211.2016.1203386.

17. Ann F. La Berge, "How the Ideology of Low Fat Conquered America," *Journal of the History of Medicine and Allied Sciences* 63,2 (2008): 139–77, doi:10.1093/jhmas/jrn001

18. Sagar B. Dugani et al., "Association of Lipid, Inflammatory, and Metabolic Biomarkers with Age at Onset for Incident Coronary Heart Disease in Women," *JAMA Cardiology* 6.4 (2021): 437–47, doi:10.1001/jamacardio.2020.7073.

19. Russell J. de Souza et al., "Intake of Saturated and Trans Unsaturated Fatty Acids and Risk of All Cause Mortality, Cardiovascular Disease, and Type 2 Diabetes: Systematic Review and Meta-Analysis of Observational Studies," *BMJ* 351 (2015), https://doi.org/10.1136/gmj.h3978; Lukas Schwingshackl and Georg Hoffmann, "Dietary Fatty Acids in the Secondary Prevention of Coronary Heart Disease: A Systematic Review, Meta-Analysis and Meta-Regression," *BMJ Open* 4.4 (2014): e004487.

20. Michele Drehmer et al., "Total and Full-Fat, but Not Low-Fat, Dairy Product Intakes Are Inversely Associated with Metabolic Syndrome in Adults," *Journal of Nutrition* 146,1 (2016): 81–89, doi:10.3945/jn.115.220699.

21. Deniz Senyilmaz-Tiebe et al., "Dietary Stearic Acid Regulates Mitochondria in Vivo in Humans," *Nature Communications* 9.1 (2018): 1–10.

22. Fahd Aziz Zarrouf et al., "Testosterone and Depression: Systematic Review and Meta-Analysis," *Journal of Psychiatric Practice* 15,4 (2009): 289–305, doi:10.1097/01.pra.0000358315.88931.fc.

23. E. K. Hämäläinen et al., "Decrease of Serum Total and Free Testosterone during a Low-Fat High-Fibre Diet," *Journal of Steroid Biochemistry* 18,3 (1983): 369–70, doi:10.1016/0022-4731(83)90117-4.

24. F. Berrino et al., "Reducing Bioavailable Sex Hormones through a Comprehensive Change in Diet: The Diet and Androgens (DIANA) Randomized Trial," *Cancer Epidemiology, Biomarkers and Prevention* 10,1 (2001): 25–33.

25. Lisa Parkinson and Russell Keast, "Oleocanthal, a Phenolic Derived from Virgin Olive Oil: A Review of the Beneficial Effects on Inflammatory Disease," *International Journal of Molecular Sciences* 15,7 (2014): 12323–34, doi:10.3390/ijms150712323.

26. Balaji Bhavadharini et al., "Association of Dairy Consumption with Metabolic Syndrome, Hypertension and Diabetes in 147,812 Individuals from 21 Countries," *BMJ Open Diabetes Research and Care* 8.1 (2020): e000826.

27. Kristin M. Nieman, Barbara D. Anderson, and Christopher J. Cifelli, "The Effects of Dairy Product and Dairy Protein Intake on Inflammation: A Systematic Review of the Literature," *Journal of the American College of Nutrition* (2020): 1–12, doi:10.1080/07315724.2020.1800532.

28. José Eduardo de Aguilar-Nascimento et al., "Early Enteral Nutrition with Whey Protein or Casein in Elderly Patients with Acute Ischemic Stroke: A

Double-Blind Randomized Trial," *Nutrition* 27,4 (2011): 440–44, doi:10.1016/j .nut.2010.02.013; Chiara Flaim et al., "Effects of a Whey Protein Supplementation on Oxidative Stress, Body Composition and Glucose Metabolism among Overweight People Affected by Diabetes Mellitus or Impaired Fasting Glucose: A Pilot Study," *Journal of Nutritional Biochemistry* 50 (2017): 95–102, doi:10.1016/j.jnutbio.2017.05.003; Gerald S. Zavorsky et al., "An Open-Label Dose-Response Study of Lymphocyte Glutathione Levels in Healthy Men and Women Receiving Pressurized Whey Protein Isolate Supplements," *International Journal of Food Sciences and Nutrition* 58,6 (2007): 429–36, doi:10.1080/09637480701253581.

29. Dagfinn Aune et al., "Nut Consumption and Risk of Cardiovascular Disease, Total Cancer, All-Cause and Cause-Specific Mortality: A Systematic Review and Dose-Response Meta-Analysis of Prospective Studies," *BMC Medicine* 14.1 (2016): 207.

30. Aune et al., "Nut Consumption."

31. Benoit Chassaing et al., "Dietary Emulsifiers Impact the Mouse Gut Microbiota Promoting Colitis and Metabolic Syndrome," *Nature* 519,7541 (2015): 92–96, doi:10.1038/nature14232.

32. Jun Lv et al., "Consumption of Spicy Foods and Total and Cause Specific Mortality: Population Based Cohort Study," *BMJ* 351 (2015): h3942, doi:10.1136/bmjh3942.

33. Hyunwoo Oh et al., "Low Salt Diet and Insulin Resistance," *Clinical Nutrition Research* 5,1 (2016): 1–6, doi:10.7762/cnr.2016.5.1.1.

34. Niels Albert Graudal et al., "Effects of Low Sodium Diet versus High Sodium Diet on Blood Pressure, Renin, Aldosterone, Catecholamines, Cholesterol, and Triglyceride," *Cochrane Database of Systematic Reviews* (November 9, 2011): doi:10.1002/14651858.CD004022.pub3.

35. Farideh Shishehbor et al., "Vinegar Consumption Can Attenuate Postprandial Glucose and Insulin Responses; a Systematic Review and Meta-Analysis of Clinical Trials," *Diabetes Research and Clinical Practice* 127 (2017): 1–9, doi:10.1016/j.diabres.2017.01.021.

36. Solaleh Sadat Khezri et al., "Beneficial Effects of Apple Cider Vinegar on Weight Management, Visceral Adiposity Index and Lipid Profile in Overweight or Obese Subjects Receiving Restricted Calorie Diet: A Randomized Clinical Trial," *Journal of Functional Foods* 43 (2018): 95–102.

37. World Health Organization, "*Nutrients in Drinking Water*," WHO/SDE /WSH/05.09, 2005.

38. Corinne E. Hall, J. P. Meyers, and Laura N. Vandenberg, "Nonmonotonic

Dose-Response Curves Occur in Dose Ranges That Are Relevant to Regulatory Decision-Making," *Dose-Response* 16,3 (2018), doi:10.1177/1559325818798282.

39. Prahlad Gupta et al., "Role of Sugar and Sugar Substitutes in Dental Caries: A Review," *ISRN Dentistry* (December 29, 2013), 2013, doi:10.1155/2013/519421.

40. Thorsten Stahl et al., "Migration of Aluminum from Food Contact Materials to Food—A Health Risk for Consumers?" *Environmental Sciences Europe* 29,17 (2017), doi: 10.1186/s12302-017-0117-x.

41. Reza Bayat Mokhtari et al., "The Role of Sulforaphane in Cancer Chemoprevention and Health Benefits: A Mini-Review," *Journal of Cell Communication and Signaling* 12,1 (2018): 91–101, doi:10.1007/s12079-017 -0401-y; Andrea Tarozzi et al., "Sulforaphane as a Potential Protective Phytochemical against Neurodegenerative Diseases," *Oxidative Medicine and Cellular Longevity* 2013 (2013), article 415078, doi:10.1155/2013/415078.

42. Hannah Ariel and John P. Cooke, "Cardiovascular Risk of Proton Pump Inhibitors," *Methodist Debakey Cardiovascular Journal* 15,3 (2019): 214–19, doi:10.14797/mdcj-15-3-214.

43. Joseph Feher, "The Stomach," in *Quantitative Human Physiology* (Elsevier Science, 2017): 785–95.

44. George Karamanolis et al., "A Glass of Water Immediately Increases Gastric pH in Healthy Subjects," *Digestive Diseases and Sciences* 53,12 (2008): 3128/32, doi:10.1007/s10620-008-0301-3.

45. Marc R. Yago et al., "Gastric Reacidification with Betaine HCl in Healthy Volunteers with Rabeprazole-Induced Hypochlorhydria," *Molecular Pharmaceutics* 10,11 (2013): 4032–27, doi:10.1021/mp4003738.

46. Gaye Tuchman and Harry Gene Levine, "New York Jews and Chinese Food: The Social Construction of an Ethnic Pattern," *Journal of Contemporary Ethnography* 22,3 (1993): 382–407, https://doi. org /10.1177/089124193022003005.

INDEX

(Page references in *italics* refer to illustrations.)

ABOUT THE AUTHOR

MAX LUGAVERE is a health and science journalist and the author of the *New York Times* bestseller *Genius Foods: Become Smarter, Happier, and More Productive While Protecting Your Brain for Life*. His sophomore book, also a bestseller, is called *The Genius Life: Heal Your Mind, Strengthen Your Body, and Become Extraordinary*. Max also hosts the number one iTunes health and wellness podcast titled *The Genius Life*.

Lugavere appears regularly on *The Dr. Oz Show*, *The Rachael Ray Show*, and *The Doctors*. He has contributed to Medscape, *Vice*, Fast Company, CNN, and the Daily Beast, has been featured on *NBC Nightly News,* the *Today* show, and in the *New York Times* and *People* magazine. He is an internationally sought-after speaker and has given talks at South by Southwest; the New York Academy of Sciences; the Biohacker Summit in Stockholm, Sweden; and other venues.

GENIUS KITCHEN. Copyright © 2022 by Max Lugavere. All rights reserved.
Printed in Canada. No part of this book may be used or reproduced
in any manner whatsoever without written permission except in the
case of brief quotations embodied in critical articles and reviews. For
information, address HarperCollins Publishers, 195 Broadway, New
York, NY 10007.

HarperCollins books may be purchased for educational, business, or
sales promotional use. For information, please email the Special Markets
Department at SPsales@harpercollins.com.

FIRST EDITION

Designed by Bonni Leon-Berman

Photographs © 2021 by Eric Wolfinger

Library of Congress Cataloging-in-Publication Data has been applied
for.

ISBN 978-0-06-302294-2

22 23 24 25 26   TC   10 9 8 7 6 5 4 3 2 1